Geographies of Science

Knowledge and Space

Volume 3

Knowledge and Space

This book series entitled "Knowledge and Space" is dedicated to topics dealing with the production, dissemination, spatial distribution, and application of knowledge. Recent work on the spatial dimension of knowledge, education, and science; learning organizations; and creative milieus has underlined the importance of spatial disparities and local contexts in the creation, legitimation, diffusion, and application of new knowledge. These studies have shown that spatial disparities in knowledge and creativity are not short-term transitional events but rather a fundamental structural element of society and the economy.

The volumes in the series on Knowledge and Space cover a broad range of topics relevant to all disciplines in the humanities and social sciences focusing on knowledge, intellectual capital, and human capital: clashes of knowledge; milieus of creativity; geographies of science; cultural memories; knowledge and the economy; learning organizations; knowledge and power; ethnic and cultural dimensions of knowledge; knowledge and action; and the spatial mobility of knowledge. These topics are analyzed and discussed by scholars from a range of disciplines, schools of thought, and academic cultures.

Knowledge and Space is the outcome of an agreement concluded by the Klaus Tschira Foundation and Springer in 2006.

Series Editor:

Peter Meusburger, Department of Geography, University of Heidelberg, Germany

Advisory Board:

Prof. Gregor Ahn, University of Heidelberg, Germany; Prof. Ariane Berthoin Antal, Wissenschaftscentrum Berlin für Sozialforschung, Germany; Prof. Joachim Funke, University of Heidelberg, Germany; Prof. Michael Heffernan, University of Nottingham, United Kingdom; Prof. Madeleine Herren-Oesch, University of Heidelberg, Germany; Prof. Friedrich Krotz, University of Erfurt, Germany; Prof. David N. Livingstone, Queen's University of Belfast, Northern Ireland, United Kingdom; Prof. Edward J. Malecki, Ohio State University, Columbus, Ohio, USA; Prof. Joseph Maran, University of Heidelberg, Germany; Prof. Gunter Senft, Max Planck Institute for Psycholinguistics, Nijmegen, The Netherlands; Prof. Wolf Singer, Max-Planck-Institute für Hirnforschung, Frankfurt am Main, Germany; Prof. Nico Stehr, Zeppelin University Friedrichshafen, Friedrichshafen, Germany; Prof. Jürg Wassmann, University of Heidelberg, Germany; Prof. Peter Weichhart, University of Vienna, Austria; Prof. Michael Welker, University of Heidelberg, Germany; Prof. Benno Werlen, University of Jena, Germany

For other titles published in this series, go to
www.springer.com/series/7568

Peter Meusburger · David N. Livingstone ·
Heike Jöns
Editors

Geographies of Science

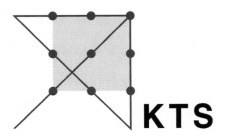

KLAUS TSCHIRA STIFTUNG
GEMEINNÜTZIGE GMBH

Editors
Peter Meusburger
Universität Heidelberg
Geographisches Institut
Berliner Str. 48
69120 Heidelberg
Germany
peter.meusburger@geog.uni-heidelberg.de

Heike Jöns
Loughborough University
Department of Geography
East Park
Loughborough, Leics.
United Kingdom LE11 3TU
h.jons@lboro.ac.uk

David N. Livingstone
Queen's University Belfast
School of Geography,
Archaeology &
  Palaeoecology
Elmwood Avenue
Belfast
United Kingdom BT7 1NN
d.livingstone@qub.ac.uk

Technical Editor:
David Antal, Berlin
Davidrantal@aol.com

ISBN 978-90-481-8610-5   e-ISBN 978-90-481-8611-2
DOI 10.1007/978-90-481-8611-2
Springer Dordrecht Heidelberg London New York

Library of Congress Control Number: 2010922067

Printed on acid-free paper

Springer is part of Springer Science+Business Media (www.springer.com)

Contents

Contributors

Dominik Collet, Dr. Seminar für Mittlere und Neuere Geschichte, Georg-August-Universität Göttingen, 37073 Göttingen, Germany, dominik.collet@phil.uni-goettingen.de

Ricardo B. Duque, Prof. Dr. Department of Sociology, Tulane University, New Orleans, LA 70118, USA, rduque@tulane.edu

Sally Eden, Dr. Department of Geography, University of Hull, Hull, HU6 7RX, UK, s.e.eden@hull.ac.uk

David M. Evans University Teacher, Department of Geography, Loughborough University, Loughborough LE11 3TU, UK, d.m.evans@lboro.ac.uk

Alexa Färber, Prof. Dr. Institut für Europäische Ethnologie, Berlin, 10117, Germany, alexa.faerber@rz.hu-berlin.de

Ryan Holifield, Prof. Dr. Department of Geography, University of Wisconsin-Milwaukee, Milwaukee, WI 53201-0413, USA, holifiel@uwm.edu

Michael Hoyler Senior Lecturer, Department of Geography, Loughborough University, Loughborough LE11 3TU, UK, m.hoyler@lboro.ac.uk

Heike Jöns, Dr. Department of Geography, Loughborough University, Loughborough LE11 3TU, UK, h.jons@lboro.ac.uk

David N. Livingstone, Prof. Dr. School of Geography, Queen's University Belfast, Belfast BT7 1NN, Northern Ireland, d.livingstone@qub.ac.uk

Peter Meusburger, Prof. Dr. Geographisches Institut, Universität Heidelberg, Heidelberg 69120, Germany, peter.meusburger@geog.uni-heidelberg.de

Thomas Schuch, M.A. Geographisches Institut, Universität Heidelberg, Heidelberg 69120, Germany, thomas.schuch@geog.uni-heidelberg.de

Wesley M. Shrum, Prof. Dr. Sociology Department, Louisiana State University, Baton Rouge, LA 70803, USA, shrum@lsu.edu

Nico Stehr, Prof. Dr. Karl Mannheim Chair for Cultural Studies, Zeppelin University, Friedrichshafen D-88045, Germany, nico.stehr@t-online.de

Peter J. Taylor, Prof. Dr. Department of Geography, Loughborough University, Loughborough LE11 3TU, UK, p.j.taylor@lboro.ac.uk

Alexander Vasudevan, Dr. School of Geography, University of Nottingham, Nottingham, NG7 2RD, UK, alexander.vasudevan@nottingham.ac.uk

Charles W.J. Withers, Prof. Dr. Institute of Geography, University of Edinburgh, Edinburgh, EH8 9XP, UK, c.w.j.withers@ed.ac.uk

Albena Yaneva, Dr. School of Environment and Development (SED), Manchester Architecture Research Centre (MARC), University of Manchester, Manchester, M13 9PL, UK, albena.yaneva@manchester.ac.uk

Marcus Antonius Ynalvez, Prof. Dr. Department of Behavioral, Applied Sciences and Criminal Justice (BASCJ), Texas A&M International University, Laredo, TX 78041, USA, mynalvez@tamiu.edu

Interdisciplinary Geographies of Science

Heike Jöns, David N. Livingstone, and Peter Meusburger

More than 2 decades into the "geographical" turn within science studies (Shapin, 1998, pp. 5–6), geographies of science are a vibrant interdisciplinary field of research. Based on exciting work by geographers, historians, sociologists, and anthropologists of science, the ideas that science has a geography and that scientific knowledge bears the marks of particular locations have themselves become accepted facts, at least within this community of scholars. Indeed, it can be argued that the meaning of scientific knowledge "takes shape in response to spatial forces at every scale of analysis—from the macropolitical geography of national regions to the microsocial geography of local cultures" (Livingstone, 2003, p. 4).

Instead of marveling at the apparent universality and "placelessness" of scientific knowledge, scholars interested in the geographies of science have focused on the specific circumstances of scientific practices and on the ways in which the travels of scientists, resources, and ideas shape the production and circulation of scientific knowledge. They also examine how and why the interpretation of certain knowledge claims may change in different times and places. The variety of research topics and approaches addressed within geographies of science is documented in a number of reviews that emphasize the long-standing mutual enrichment of research carried out in geography and other fields that contribute to interdisciplinary science studies (Finnegan, 2008; Livingstone, 1995, 2003; Meusburger, 2008; Naylor, 2005; Powell, 2007; Shapin, 1995, 1998; Withers, 2002).

H. Jöns (✉)
Department of Geography, Loughborough University, Loughborough, LE11 3TU, UK
e-mail: h.jons@lboro.ac.uk

D.N. Livingstone
School of Geography, Queen's University Belfast, Belfast, BT7 1NN, Northern Ireland
e-mail: d.livingstone@qub.ac.uk

P. Meusburger
Geographisches Institut, Universität Heidelberg, Heidelberg, 69120, Germany
e-mail: peter.meusburger@geog.uni-heidelberg.de

A defining moment in this reciprocal relationship is captured by Livingstone's (1995) outline of a "historical geography of science" that explores the contributions Michel Foucault, Clifford Geertz, Anthony Giddens, Donna Haraway, Bruno Latour, Edward Said, and others have made to the conceptualization of a distinctively geographical interest in scientific knowledge and practice. Shapin (1998) responded that "[s]tudents of science owe much to geographers and it is flattering to learn that Livingstone thinks that historians of geography might possibly learn something from us. If so, it is mainly through showing some of the possibilities inherent in geographical work" (p. 6). This conversation between geography and science studies has continued to flourish ever since. It has not only produced a series of commentaries on the value of social constructivism and actor-network thought for the geographies of science (e.g., Barnes, 1998, 2001; Bravo, 1999; Demeritt, 1996, 2006; Harris, 1998; Jöns, 2006) and for human geography more generally (e.g., Bingham & Thrift, 2000; Murdoch, 1997) but has also inspired substantial monographs (e.g., Ash & Cohendet, 2003; Driver, 2001; Livingstone, 2002, 2003; Whatmore, 2002; Withers, 2001), comprehensive anthologies (e.g., Simões, Carneiro, & Diogo, 2003; Smith & Agar, 1998), and a number of special journal issues (e.g., Anderson, Kearnes, & Doubleday, 2007; Castree & Nash, 2006; Naylor, 2005; Philo & Pickstone, 2009; Roe & Greenhough, 2006). Among the most recent outcomes are the seminar series and online reader entitled *Locating Technoscience* (UCL, 2008) and the "Knowledge and Space" symposia and book series, of which the present collection of essays is the third volume.

Aiming to further advance interdisciplinary geographies of science through conversations between scholars working in different academic fields, this volume explores the benefits of a geographical perspective on scientific knowledge and practice from the perspective of geographers, sociologists, historians, anthropologists, and scholars of architecture. A comparison of their contributions both discloses how different disciplinary settings exert their influence on framing research designs in distinct ways and indicates a common concern for the spatial relations of scientific knowledge and practice. The book presents a balance of historical and contemporary case studies, with most of the essays centering on European practices. However, some of the chapters provide global perspectives, whereas others deal with African practices and American indigenous knowledges. Keeping in mind that one of the most significant insights into the spatiality of knowledge production is the partiality of all knowledge claims (Haraway, 1988), we note that the following peer-reviewed essays inevitably provide very specific perspectives on the geographies of science. These chapters add to a growing body of work yet also raise important questions for future research.

This volume stresses four main topics, each of which is represented in a corresponding section. The first, "Comparative Approaches to Scientific Knowledge," gives two fairly general accounts—one by historical geographer David N. Livingstone (chapter "Landscapes of Knowledge") and the other by sociologist Nico Stehr (chapter "Global Knowledge?"). Aiming to further develop the agenda of geographical science studies, Livingstone delineates the overall context for this set of essays. He begins by reviewing ways in which space has become a central organizing principle for examining the production, circulation, and consumption of scientific knowledge, stating that scientific sites and spaces, the movement and transformation of knowledge, and scientific regions ranging from the provincial to the continental have been significant foci of research. Livingstone then discusses how geographies of science have challenged long-standing polarities such as the natural and the social, the local and the global, and the scientific and the political. He also outlines the benefits of bringing materialities of science to center stage, pointing out that scientific knowledge resides in bodies, buildings, and other physical objects. Lastly, Livingstone elaborates on four spatial themes for future research: the agency of landscape, political ecology, print culture, and speech space.

Stehr approaches the spatiality of knowledge from a slightly different angle by discussing the idea of "global worlds of knowledge." Interestingly, however, he arrives at a conclusion not altogether different from Livingstone's notion of geographically diverse landscapes of knowledge. Stehr distinguishes between the horizontal integration of knowledge (meaning the proliferation of sites of knowledge production and consumption) and the vertical integration of knowledge (meaning the relationship between expert knowledge and everyday knowledge across social worlds). On this basis he reasons that globalizing worlds of knowledge may partially exist as "normative speculations, by decree, as a thought experiment, or as a business plan" but that the challenges and constraints are far too large for anything like a comprehensive global world of knowledge to emerge.

This book's second part, "Mobilities and Centers," is written by geographers. It draws attention to the circulatory spaces of science by examining how transient and more permanent moves of scientists and scholars between different sites of academic knowledge production have contributed to the formation of scientific centers. Studying the career paths of eminent scientists in Europe, Peter J. Taylor, Michael Hoyler, and David M. Evans (chapter "A Geohistorical Study of "The Rise of Modern Science": Mapping Scientific Practice Through Urban Networks, 1500–1900") identify the shifting geographies of European knowledge centers and their networks from the sixteenth to the nineteenth centuries. Interpreting scientific practice as a core-producing process in Wallerstein's (2004) modern world-system, the authors assert that studying the work places and career moves of scientists yields information about the two types of social space identified by Castells (1996): spaces of places and spaces of flows. The resulting geohistorical patterns of European knowledge nodes and networks provide a unique macroperspective on the "rise of modern science" that simultaneously offers an argument about why so many European scientific centers did not become major cities.

Peter Meusburger and Thomas Schuch (chapter "From Mediocrity and Existential Crisis to Scientific Excellence: Heidelberg University Between 1803 and 1932"), too, use data on the career mobility of scientists and scholars, tracking the rise of Heidelberg University to the ranks of internationally renowned research universities in the late nineteenth and early twentieth centuries. They illustrate how that trajectory is mirrored in the changing social background of Heidelberg's professors; the age at which they reached different career stages; and the growing diversity of the places in which they were born, received their doctoral and postdoctoral degrees, and became professors. The authors hold that Heidelberg University's favorable working environment in the nineteenth century—due to effective reorganization and financial support by the state, broad university autonomy, freedom of thought, and an accommodating political climate—permitted increasingly selective recruitment policies targeting renowned professors at the peak of their careers. Consequently, Heidelberg's full professors were often highly mobile individuals who had worked in a variety of cultural environments, a situation that reveals how the openness to drawing faculty from geographically diverse places nurtured the formation of an important scientific center.

Heike Jöns (chapter "Academic Travel from Cambridge University and the Formation of Centers of Knowledge, 1885–1954") examines a more transient circulation of academics by looking at the ways in which a growing emphasis on academic travel for purposes of research, visiting appointments, lecturing, conferences, and consulting contributed to transforming Cambridge University into a modern research university. She conceptualizes circular academic mobility as a twofold mobilization process in Latourian "centers of calculation," namely, the home institutions and the host institutions. This perspective sheds light on how, from the 1890s onward, the temporary recruitment of Cambridge expertise in the United States—mainly through visiting appointments and lecture tours—gradually turned American universities into new global scientific centers and fostered the development of an Anglo-American academic hegemony in the twentieth century.

The third part of this volume, "Designing Knowledge Spaces," discusses four attempts to create distinct spaces for knowledge production and consumption. Historian Dominik Collet (chapter "Big Sciences, Open Networks, and Global Collecting in Early Museums") examines the endeavor by the fellows of London's Royal Society to establish their own museum for research and the display of specimens in the second half of the seventeenth century. Interrogating Lux and Cook's (1998) hypothesis that weak, but flexible, "open networks" were crucial for scientific progress in early modern times, he shows why the fellows' truly global network of correspondence supplied a number of objects regarded as "exotic curiosities" but produced scientifically rather unsatisfying results. The author contends that unreliable, uncooperative colonial contacts and the disparate information of poor quality that often reached London via routes different from those traveled by the material objects themselves made it impossible to gather the contextual information required for serious scientific research. Without such an intact Latourian "circulating reference" between the museum's specimens and their places of origin, the Royal Society's widespread open networks failed to spur scientific progress and thus restricted the collection's function to the preservation and presentation of curiosities.

Albena Yaneva (chapter "Is the Atrium More Important than the Lab? Designer Buildings for New Cultures of Creativity"), a sociologist and ethnographer working in the field of architectural studies, draws attention to contemporary buildings conceived for scientific practice. She critically examines recent efforts to design attractive atria intended to facilitate social interaction and generate creative encounters beyond confined laboratory spaces. Starting with a discussion of design principles for recent laboratory spaces, she maintains that the emphasis on the atrium is a response to the challenge of enhancing the potential for collaborative research and networking between human and nonhuman actors across disciplinary boundaries. Her examples explain how the atrium became an "important interactive space"; a "social core" with multiple bridging functions; a transdisciplinary "mixing chamber" of researchers, objects, and ideas; even "a complex knot of a quasi-urban network" in city-shaped buildings that correspond to the complex research tasks at hand. Although not all of the innovative designs have been popular with scientists, the realities arising from the discussed projects for improving academic working environments starkly contrast those of most such work places, which are too often characterized by the much less socially conducive campus architecture of the 1960s and 1970s.

The contribution by sociologists Wesley Shrum, Ricardo B. Duque, and Marcus Antonius Ynalvez (chapter "Outer Space of Science: A Video Ethnography of Reagency in Ghana") suggests, however, that some scholars and scientists may find architecture's inspirational qualities less important than what they consider to be basic, functioning e-mail and Internet infrastructure at the university. This impression is conveyed by the authors' hitherto unsuccessful attempt to facilitate Internet connectivity in a Ghanaian research institute. The story of this project, which turned attention to what the authors called the "outer space of science," was presented in Heidelberg in the form of a video ethnography ensuing from 2 years of work. For this collection of essays, Shrum and his colleagues retold the basic story line and critically reflected upon the ways in which the original video ethnography was received by the audience in Heidelberg. Originally, the authors had obtained US National Science Foundation funds to examine the use of the Internet and its effects on social networks of scientists in Africa and India. But the sites of interest in Ghana lacked Internet connectivity, so the funds were rededicated in order to provide for this basic condition of the primary research interest.

The study shows that new information and communication technologies, depending on the quality of the services, seem to be of ambiguous value for some academics working in Ghana. It also reveals that external funding from the US team was identified by other Ghanaians—quite independently of the American project members—as a means for making money rather than for making progress toward Internet connectivity. The failure of the project frustrated the authors but also gave rise to a wonderful academic friendship with a Ghanaian lecturer in sociology. This outcome highlights two points: (a) the contingency of transnational academic exchange and (b) the fact that the spaces of science often taken for granted by academics are in fact very fragile, difficult to achieve and sustain, and geographically very concentrated.

In this section's final paper, which is based on ethnographic fieldwork in Rabat and Hanover, ethnographer Alexa Färber (chapter "The Making of Geographies of Knowledge at World's Fairs: Morocco at Expo 2000 in Hanover") also discusses relationships between the global South and the global North. However, she explores the design of space for knowledge consumption by analyzing the ways in which a team of former politicians, diplomats, civil servants, government advisors, architects, and academics (including two geographers) constructed Morocco's representation at EXPO 2000 in Hanover. The author elaborates the reasons why the committee members responsible for the country's "representational work" did not address the realities of modern Morocco with technological media but instead undertook to anticipate the visitors' expectations as potential tourists and therefore concentrated on displaying cultural heritage through "artisans, folklore, and artifacts." Färber argues that drawing on the oriental and world-fair aesthetic archives in order to live up to "Western fantasy" not only rendered the "new smartness" of Morocco's knowledge society invisible but reproduced knowledge divides between a smart global North and an ignorant global South. Her article therefore demonstrates that the public influenced the production of geographical knowledge so much that it shaped the reasoning of the academic experts involved in designing Morocco's self-presentation on an international stage.

In a series of articles written by geographers, this book's final part, "Science and the Public," further explores important interactions between these two realms. Using newspaper reports and other written sources dating from the period 1845 to 1939, Charles W. J. Withers (chapter "Geographies of Science and Public Understanding? Exploring the Reception of the British Association for the Advancement of Science in Britain and in Ireland, c.1845–1939") provides a historical perspective on how the public received the peripatetic annual scientific meetings of the British Association for the Advancement of Science. He illustrates how the attendance at and reaction to presentations varied considerably, depending on complex issues such as the differences between the BAAS's different thematic sections, "popular" and "scientific" presentations, more and less prolific speakers, those using lantern slides and those who did not visually support their stories, the local audiences' perceptions of the Association's objectives, and gender. The author holds that the interaction with the public at the BAAS meetings was often more akin to "participating in a civic social gathering" than to a genuine interest in the content of science. But he also asserts that it would be misleading to speak of a homogenous public, for the documented variations in attendance, reception, and understanding tend to bear out the concept of a historically and geographically contingent relationship between heterogeneous sciences and multiple publics.

Alexander Vasudevan (chapter "Testing Times: Experimental Counter-Conduct in Interwar Germany") takes the reader to Weimar Germany, where he explores the relationship between modernist art experiments and the experimental life sciences, particularly "psychiatric science." The first of his two case studies shows how Berlin Dada used the stylistic device of montage to transport the issue of war neurosis "from the trenches and clinics to the sites and venues of postwar metropolitan culture" and to address the physical and mental consequences of "the shock of urban industrial modernity." The second case study investigates the ways in which psychotechnics widely employed to raise workplace efficiency were used in Brechtian epic theater to transform the audience from test subjects in everyday life into informed experts in the theater. Although Brecht's 1931 production of *Mann ist Mann* was rather critically received by the audiences, experimental psychiatry "furnished Berlin Dada and Brechtian epic theater with a new repertoire of performance styles and representational techniques," creating an "alternative experimental program" that contested mainstream German psychiatry. His article thus suggests that both performative art and political theater offer "a critical perspective on the extension of the experimental into nonscientific zones."

Sally Eden's essay (chapter "NGOs, the Science-Lay Dichotomy, and Hybrid Spaces of Environmental Knowledge") looks at contemporary interactions between science and the public by exploring the ways in which NGOs engage with science when advancing their agendas for environmental reform. Adopting Gieryn's concept of "boundary work" (Gieryn, 1983), she maintains that NGOs complicate the simplistic dichotomy between scientific experts and "a supposedly lay public" in many ways, such as by recruiting more and more researchers with postgraduate degrees and by enrolling scientists who support their moral agenda. Some NGOs go beyond bridging work by deliberately creating hybrid spaces of "heterogeneous knowledge practices" for their purposes, as when they draw on an international panel of scientific experts and environmental practitioners. Eden suggests that these hybridizations are not only variously successful but also highly specific in time and space. A British example relates to the Forest Stewardship Council's national standards governing the acceptable use and types of pesticides, the revision of which every 5 years is based on the latest research findings. The boundaries of these hybrid lay–expert knowledge spaces appear to be much more dynamic, fuzzy, and blurred than those of the modernist dichotomies between science and politics that they undermine. One can thus regard the former kind of boundaries as more flexible than the latter kind for both challenging and building alliances with science in environmental policy debates.

The final essay of this book, written by Ryan Holifield (chapter "Regulatory Science and Risk Assessment in Indian Country: Taking Tribal Publics into Account"), begins by calling attention to the contested nature of such hybrid knowledge spaces. Specifically, he points out that the practices of risk assessment as conducted by the US Environmental Protection Agency (EPA) have been severely criticized by both regulated industries (for being overly protective) and environmental activists (for not being protective enough). He focuses on regulatory science as opposed to academic science in his discussion of how the debates about localizing the procedures of human health risk assessment in Indian Country in the United States have developed since the 1980s. Drawing on Latour's concept of collectives of humans and nonhumans, which corresponds well to tribal, or nonmodern, traditional worldviews of integrated human communities and nonhuman environments, Holifield explains why tribal traditional lifeways escape the EPA's standard risk assessment procedures attuned to typical suburban populations. In other words, they require attention to the voices of locally distinctive publics as "nations within." He argues that regulatory science must engage with multiple publics as well as human and nonhuman collectives in order to secure credibility and legitimacy.

In conclusion, the pluralization and multiplicity of science and the public, scientific centers and designs, as well as concepts and approaches run as a central theme through the main sections of this book, highlighting the spatial and temporal complexity and contingency of past, present, and future interdisciplinary geographies of science. Our abiding thanks go to the Klaus Tschira Foundation for generously funding the symposia and book series on Knowledge and Space and for thereby making this productive interdisciplinary encounter possible. We are also grateful to Edgar Wunder, Christiane Marxhausen, and their team in Heidelberg for organizing the symposia and assisting with the production of this book and to David Antal for his thoughtful contributions as the technical editor of this book series.

References

Anderson, B., Kearnes, M., & Doubleday, R. (2007). Geographies of nano-technoscience. *Area*, *39*, 139–142.

Ash, A., & Cohendet, P. (2003). *Architectures of knowledge: Firms, capabilities and communities*. Oxford: Oxford University Press.

Barnes, T. J. (1998). A history of regression: Actors, networks, machines, and numbers. *Environment and Planning A*, *30*, 203–223.

Barnes, T. J. (2001). 'In the beginning was economic geography': A science studies approach to disciplinary history. *Progress in Human Geography*, *25*, 455–478.

Bingham, N., & Thrift, N. (2000). Some new instructions for travellers: The geography of Bruno Latour and Michel Serres. In M. Crang & N. Thrift (Eds.), *Thinking space* (pp. 281–301). London: Routledge.

Bravo, M. T. (1999). Ethnographic navigation and the geographical gift. In D. N. Livingstone & C. W. J. Withers (Eds.), *Geography and enlightenment* (pp. 199–235). Chicago: University of Chicago Press.

Castells, M. (1996). *The information age: Vol. 1. The rise of the network society*. Oxford, England: Blackwell.

Castree, N., & Nash, C. (2006). Editorial: Posthumanist geographies. *Social and Cultural Geography*, *7*, 501–504.

Demeritt, D. (1996). Social theory and the reconstruction of science and geography. *Transactions of the Institute of British Geographers, New Series*, *21*, 483–503.

Demeritt, D. (2006). Science studies, climate change and the prospects for constructivist critique. *Economy and Society*, *35*, 453–479.

Driver, F. (2001). *Geography militant: Cultures of exploration and empire*. Oxford, England: Blackwell.

Finnegan, D. (2008). The spatial turn: Geographical approaches in the history of science. *Journal of the History of Biology*, *41*, 369–388.

Gieryn, T. F. (1983). Boundary-work and the demarcation of science from non-science: Strains and interests in professional ideologies of scientists. *American Sociological Review*, *48*, 781–795.

Haraway, D. (1988). Situated knowledges: The science question in feminism and the privilege of partial perspective. In M. Biagioli (Ed.), *The science studies reader* (pp. 172–188). New York: Routledge.

Harris, S. J. (1998). Long-distance corporations, big sciences, and the geography of knowledge. *Configurations*, *6*, 269–304.

Jöns, H. (2006). Dynamic hybrids and the geographies of technoscience: Discussing conceptual resources beyond the human/non-human binary. *Social and Cultural Geography*, *7*, 559–580.

Livingstone, D. N. (1995). The spaces of knowledge: Contributions towards a historical geography of science. *Environment and Planning D: Society and Space*, *13*, 5–34.

Livingstone, D. N. (2002). *Science, space and hermeneutics* (Hettner Lectures Vol. 5). Heidelberg: Department of Geography.

Livingstone, D. N. (2003). *Putting science in its place: Geographies of scientific knowledge*. Chicago: University of Chicago Press.

Lux, D., & Cook, H. (1998). Closed circles or open networks? Communicating at a distance during the scientific revolution. *History of Science*, *36*, 179–211.

Meusburger, P. (2008). The nexus of knowledge and space. In P. Meusburger (Series Ed.) & P. Meusburger, M. Welker, & E. Wunder (Eds.) *Knowledge and space: Vol. 1. Clashes of knowledge: Orthodoxies and heterodoxies in science and religion* (pp. 35–90). Dordrecht: Springer.

Murdoch, J. (1997). Towards a geography of heterogeneous associations. *Progress in Human Geography, 21*, 321–337.

Naylor, S. (2005). Introduction: Historical geographies of science: Places, contexts, cartographies. *British Journal for the History of Science, 38*, 1–12.

Philo, C., & Pickstone, J. (2009). Unpromising configurations: Towards local historical geographies of psychiatry. *Health and Place, 15*, 649–656.

Powell, R. C. (2007). Geographies of science: Histories, localities, practices, futures. *Progress in Human Geography, 31*, 309–329.

Roe, E. J., & Greenhough, B. (2006). Editorial: Towards a geography of bodily technologies. *Environment and Planning A, 38*, 416–422.

Shapin, S. (1995). Here and everywhere: Sociology of scientific knowledge. *Annual Review of Sociology, 21*, 289–321.

Shapin, S. (1998). Placing the view from nowhere: Historical and sociological problems in the location of science. *Transactions of the Institute of British Geographers, New Series, 23*, 5–12.

Simões, A., Carneiro, A., & Diogo, M. P. (2003). *Travels of learning: A geography of science in Europe*. Dordrecht: Kluwer Academic Publishers.

Smith, C., & Agar, J. (Eds.). (1998). *Making space for science: Territorial themes in the shaping of knowledge*. Basingstoke: Macmillan.

UCL (University College London) (Ed.). (2008). *Locating Technoscience On-line Reader*. Retrieved May 29, 2009, from UK Internet: http://www.ucl.ac.uk/sts/locating-technoscience Version 1.0

Wallerstein, I. (2004). *World-systems analysis: An introduction*. Durham, NC: Duke University Press.

Whatmore, S. (2002). *Hybrid geographies: Natures cultures spaces*. London: Sage.

Withers, C. W. J. (2001). *Geography, science and national identity: Scotland since 1520*. Cambridge: Cambridge University Press.

Withers, C. W. J. (2002). The geography of scientific knowledge. In N. A. Rupke (Ed.), *Göttingen and the development of the natural sciences* (pp. 9–18). Göttingen: Wallstein.

Part I
Comparative Approaches

Landscapes of Knowledge

David N. Livingstone

Space is rapidly becoming a central organizing principle for making sense of scientific knowledge. The recently published third volume of *The Cambridge History of Science*, which deals with "Early Modern Science" (Park & Daston, 2006b), is indicative. Its editors have chosen to devote nine chapters to such subjects as markets, piazzas, and villages; houses and households; libraries and lecture halls; courts and academies; anatomy theaters and botanical gardens; and coffeehouses and print shops. All are interrogated as critical sites of scientific knowledge. This emphasis, standing in marked contrast to earlier heroic narratives of scientific progress and great-name history, enables the editors to speak of the ways in which what they call the "geography of changes in natural knowledge closely tracked that of religious, military, and economic developments" (Park & Daston, 2006a, p. 7). And it raises profound questions that go beyond the mere charting of place-based activities. Eamon (2006), for example, notes that the emergence of the marketplace "as a site of natural knowledge" where goldsmiths, herbalists, apothecaries, dyers, and many other craftsmen procured fruit, leaves, seeds, ointments, and other natural objects "signaled important shifts in the definition of knowledge and of who might qualify as natural knowers. It also raised questions . . . about whose knowledge was considered valid and authoritative" (p. 207). Cooper (2006) uses her interrogation of the home as a space of knowledge to show how "natural inquiry in early modern Europe . . . often constituted a *family* project" (p. 225). Household recipe books are just as useful as scribbled laboratory notes in gleaning a sense of domestic space as a setting for scientific knowledge. English women of means, for example, sometimes had stills and alembics in their kitchens which they used to tinker "with and write down medical recipes" (Cooper, 2006, p. 227). For his part, Johns (2006) begins his analysis of coffeehouses and print shops with the arresting claim that "experimental philosophy came to prominence on a wave of coffee" (p. 320).

D.N. Livingstone (✉)
School of Geography, Queen's University Belfast, Belfast BT7 1NN, Northern Ireland
e-mail: d.livingstone@qub.ac.uk

P. Meusburger et al. (eds.), *Geographies of Science*, Knowledge and Space 3,
DOI 10.1007/978-90-481-8611-2_1, © Springer Science+Business Media B.V. 2010

This particular probing of the sites of scientific knowledge production during the early modern period is only one of the recent expressions of a more general geographical turn in science studies. Withers's (2007) remarkable geographical scrutiny of the so-called Enlightenment is another. The spatiality of knowledge has thus become a focal point of conversation amongst a range of historians, sociologists, geographers, and anthropologists interested in the nature of scientific culture. Taken in the round, it is an enterprise operating with the conviction that science "is indelibly marked by the local and the spatial circumstances of its making; that scientific knowledge is embodied, residing in people and in such material objects as books and instruments ... and, finally, that scientific knowledge is made by and through mundane—and locally varying—modes of social and cultural interaction" (Shapin, 1998, p. 6). Perspectives have differed, of course, and strategies for elucidating just how space matters have been far from uniform. Some scholars have been motivated by the ethnographic lure of thick description; some have retained the epistemological preoccupations of traditional philosophy of science but have sought to spatialize the conventional distinction between the context of discovery and the context of justification; others have dwelt on the reciprocal connections between the practices of science and the production of space.

Mapping Scientific Space

In pursuing this enterprise, a variety of ways of thinking about the spatiality of scientific knowledge and practice have emerged (Meusburger, 2000). Some of them have discriminated between the production and consumption, or construction and reception, phases of the scientific knowledge circuit. What animates this line of inquiry is the recognition that very specific kinds of spaces have to be made for the conduct of scientific inquiry. They include sites like laboratories and museums, observatories and dissecting rooms, survey ships and census bureaus, and botanical and zoological gardens. But they also embrace "natural" locations that are delimited in certain kinds of ways so as to be constituted as field sites. In each case there are protocols for the management of the space, and various mechanisms are installed to police the site and control its human occupation. In recent years, a good deal of attention has been directed to these locations and to how their microgeographies have shaped the practices that go on in these cognitively privileged sites (see Livingstone, 2003a). And there are, of course, other less dedicated spaces where scientific knowledge has been generated, including public houses, royal courts, cathedrals, tents, stock farms, and specific cities—like Chicago (Gieryn, 2006). On the consumption side of the equation, attention has been directed to how knowledge moves from its point of construction and out into the world of general intellectual commerce. A whole suite of proposals has been advanced to get a handle on the processes involved. One of the most popular has been Bruno Latour's account in which centers of calculation and immutable mobiles have held pride of place. Just how distributed knowledge and information is brought together in dedicated sites and reassembled has been at the forefront of his inquiries. Indeed, the interventions of Latour and others scrutinizing how knowledge travels and transforms have rendered troublesome the seemingly clear boundary between production and consumption, and several writers have called for the demolition of that convenient taxonomy. The reason is that knowledge is produced in the moment of encounter with new theory as it is shaped, taken up, and put to use in different intellectual and social spaces. Secord (2004) has recently given voice to this sentiment by his proposal to shift focus and think about knowledge-making as a form of communicative action (p. 661) and thereby "recognize that questions of 'what' is being said can be answered only through a simultaneous understanding of 'how', 'where', 'when', and 'for whom'" (p. 664).

Another way of thinking about the geography of science is to approach the problem through scales of spatial analysis. On one scale, it makes sense to look at very specific venues of the sort I have itemized already. On another scale, one might profitably inquire into the significance of regions (ranging from the provincial to the continental) in the conduct of scientific enterprises. Just how a scientific culture is shaped by its regional setting is likely to be a reflection of forces such as local styles of patronage, pedagogic traditions, circuits of communication, networks of social organization, and expressions of religious devotion. Historically, scientific subcultures have taken form in response to the dictates of urban politics and industrial pollution, the demands of civic pride, and radical protest. The traffic should not be thought of as one-way. It is not just that scientific practice is molded by regional setting; it is also that regional identity itself has often been conditioned by scientific projects. Applied astronomy, precision mapping, resource inventory, and geodetic survey are just a few of the scientific practices that states have mobilized for the purpose of defining the bounds of its territory and providing a register of its natural assets. Such activities at once impose rational order on the seeming chaos of nature, deliver to governments a sense of territorial coherence, and supply servants of the state with geographical data essential for fixing tax, stimulating economic growth, exploiting resources, and maintaining military defense.

A third means of thinking in a structured way about the location of science is to discriminate between different kinds of scientific spaces—such as spaces of experiment, spaces of exhibition, and spaces of expedition—and choreograph their differing geographies. Consider how the natural world is differently encountered in these different venues. In experimental laboratories, nature is subject to a variety of manipulations. In exhibitionary arenas—such as botanical or zoological gardens, or museums—the overriding concern is to arrange plants or animals or objects through some classification system and exhibit them appropriately, to put them in their place as it were. In expeditionary settings, the aim is to encounter unmanipulated nature, to literally keep it in site. Of course, there are problems with this tripartite arrangement, not the least of which is that they overlap and intersect in all sorts of promiscuous ways. As Kohler (2002), for example, has compellingly shown, the distinction between laboratory and field does not work well for major strands of biology during the early twentieth century. His inquiry into what he terms the "landscapes and labscapes" of science subverts the standard distinction. What he calls the "lab-field frontier" that "in 1900 was a defensive boundary demarcating the different worlds of field and laboratory biology" had changed by midcentury into "a broad zone of scientific subcultures that were neither pure lab nor pure field but mixtures of the two" (p. 293; for Germany, see Cittadino, 1990). It is in this third kind of scientific space that the labscape emerges as a critical site of biological inquiry and Kohler (2002) conducts a "transect" (p. 293) of this terrain elucidating the shifts in the zone's practices of place. And it is no surprise that Kohler resorts to geographical vernacular to describe his self-appointed task. He thus speaks of the "cultural geography of border biology" (p. 294), the value of a "cultural-geographic approach" (p. 295), and the blurred spaces between nature and artifice as "a patchy cultural landscape" (Kohler, 2002, p. 308).

My aim so far has not been to defend any particular systematization but rather to advertise just a few of the ways in which the geographies of science have been schematized. Various surveys of this general terrain now exist, so further review would be redundant in this chapter (see Finnegan, 2008; Naylor, 2005; Powell, 2007; Withers, 2002).

Recurrent Signals

A number of themes snaking their way through these endeavors can be extracted as focal points of recent thinking about the spatiality of science. I want now to consider them in a fairly impressionistic way. For convenience, I will briefly tackle them under two labels: *polarities* and *materialities*.

One collective consequence of the turn to the spaces of science has been to render problematic a series of polarities that have frequently been taken for granted. Let me just name three: the natural and the social; the local and the global; and the scientific and the political. First, the distinction between the natural and the social has been disturbed in different ways, not least by the recognition that in so many scientific arenas the "natural" is stage-managed by human artifice. Collins (1988), writing on the modern public experiments of the nuclear industry, observes that the demonstrations are effective precisely because the smooth public display conceals "the untidy craft" of the scientist; demonstrations work—as he superbly catches the character of the circuit from private to public—by "caging Nature's caprices in thick walls of faultless display" (p. 728). Whether in the laboratory, in the field, or in the museum, the natural is constituted by the social. Reflecting even for a moment on terms like *species, race, matter,* even *nature* itself, not to mention *selfish genes,* brings one face to face with labels that have been freighted with cultural baggage.

Second, the standard juxtaposition of the local and the global has been undermined by work on the processes of knowledge-making. It is not that universal knowledge is simply true, whereas local knowledge is at best partial or at worst pathological. Rather, the question is how locally generated scientific knowledge achieves universality (Shapin, 1998). What are the mechanisms by which it spreads—and spreads unevenly—across space and time? What role is played by the standardization of measurement, the calibration of equipment, and the disciplining of observers in global circulation? The insight that the global is local at every point has troubled the ease with which local and global may be disaggregated.

Third, the comfortable distinction between the scientific and the political has been progressively undermined in critically significant ways. Whether one is inspecting the imperial vocabulary of Darwinian biogeography (Browne, 1996); the reliance of Victorian scientific travelers on the infrastructures of colonialism (Camerini, 1996); the operations of pharmaceutical corporations in the global trade in genetic resources (Parry, 2004, 2006); the colonial networks that facilitated the transfer of botanical specimens to and from metropolitan centers like Kew Gardens (Schiebinger & Swan, 2005); the ways in which ethologists like Karl Vogt read animal behavior through the lens of the emergence of the nation-state, the decline of monarchy, and the rise of republicanism (Rupke, 2009); or the role of different regulatory regimes for the conduct of clinical trials—such as in Cuba—in the evolution of the biotech industry (Reid-Henry, in press), one sees plainly what might be called the geopolitics of scientific knowledge and practice.

Another suite of concerns grounding scientific knowledge in the places of its making congregates around what, for convenience, I call materialities. What I am after in this context is the move away from thinking about scientific knowledge as free-floating and transcendental to thinking about it in a way that roots such undertakings in material entities—like bodies, buildings, and other physical objects. In contrast to the image of scientific knowledge as disembodied, abstracted truth, a set of arguments is now in place emphasizing the corporeality of knowledge and its incarnation in human subjects. In part, this development builds on the insights of Michael Polanyi, who perceived scientific instruments as extensions of the body in the acquisition of knowledge. As he observed: "Our body is always in use as the basic instrument of our intellectual and practical control over our surroundings" (Polanyi, 1959, p. 31). In part, it draws on work directing attention to the human body as itself a site of knowledge acquisition. Alexander von Humboldt, for example, used his own body as a recording instrument on his expedition to South America between 1799 and 1804, when he applied electrodes to himself in the attempt to ascertain the effects of an electric current. Given the fact that bodies are resolutely located in space, there are grounds for suggesting that scientific knowledge is always positioned knowledge; rationality, always situated rationality; inquiry, always located inquiry (Lawrence & Shapin, 1988).

Scientific knowledge, of course, is not just incarnated in bodies; it is also located in buildings. And the role of scientific buildings in the building of science has attracted considerable interest (Gieryn, 2002). There is, for example, compelling work on such themes as the struggles over the arrangement of exhibition space, the Gothic revival style of architecture used for Victorian natural history museums as part of a cultural struggle to professionalize science, the layout of laboratories to facilitate preferred social interaction. Whatever the particulars, these concerns have served to underscore the importance of place in what might be called the architecture of science (Galison & Thompson, 1999).

Finally, the tracing out of the geographies of science has fostered increased interest in another set of material entities—objects. Whether it is human tissue, fossil specimens, geological samples, cultural artifacts, or plant species, scientific knowledge is bound up with objects that move around the world from site to site. The cartography of these dynamics at once maps physical *and* conceptual movement. Hill (2006a), for example, has traced the shifting meaning of medical objects gathered by Henry Solomon Wellcome as they traveled from their point of origin in local indigenous cultures via a medical collection to the Fowler Museum in Los Angeles. In transit, they shifted from being everyday articles to exemplars of ethnological history to works of primitive art (Hill, 2006a, 2006b). Similarly, Dritsas (2005) followed the trail of the freshwater mussel shells collected by John Kirk during the Zambezi Expedition of 1858–1864 from southeastern Africa to Philadelphia via London and thereby disclosed how the movement of these objects was critical to the warranting of zoological knowledge. This geographical trajectory was both a physical migration *and* a conceptual journey from "the farthest empirical and geographical peripheries into the metropolitan knowledge system" (Dritsas, 2005, p. 49).

Having noted some of these recurring themes in recent work on the spatiality of science, I shall now try to move the debate forward a little more by reflecting on four particular spatial themes that, in my view, might be more prominently integrated into the geography-of-knowledge project than they are at the moment.

Landscape Agency

As a consequence of my expressed unease at the deterministic cast of Dorn's *Geography of Science* (1991), some commentators have concluded that the geography of knowledge that I have promulgated is too resolutely culturalist and consequently fails to accord any role to the natural world in the shaping of cognitive claims about it. Dorn's account, let me remind you, very largely congregated around a Wittfogel-style narrative that attributed the development of science to the effects of those societies requiring hydraulic management and, thus, technoscientific initiative. To explain the development of science, Dorn looked to "soil, climate, hydrology, and topographical relief, and to demographic fluctuations, latitude, and the differences between sown fields, steppe, and desert" (p. xi). Not surprisingly he found that the writings of the U.S. geographer Ellen Churchill Semple still "remained fresh" (p. xii) and insisted that Ellsworth Huntington's environmental determinist "thesis was never really refuted" (p. xix). Now it is indeed the case that I remain profoundly uneasy with this form of ecological constructivism. But this stance should not be taken to mean that I believe physical landscapes exert *no* influence on the production of scientific knowledge. Keeling (2007) contends that my use of place has been overdetermined by a focus on the "cultural, social, and even textual spaces of scientific endeavour" (p. 405). "This emphasis on the cultural and social geographies of knowledge production," Keeling goes on, "privileges the representations and practices of scientists, devoting rather less attention to the natural spaces and phenomena which they study and engage" (p. 405). There may well be something to this claim, but I do want to allow for the agency of landscape in the production of at least some forms of scientific knowledge.

Of course, before pushing the matter of the agency of nature very far, one needs to consider in a much more sustained way the nature of agency. Traditional conceptions of agency have tended to converge on the critical importance of intentionality as the motor of history. Essentially, this formulation has restricted agency to the operation of the human. Resisting this understanding, various efforts have been made to liberate agency from human captivity. Latour (1999), for example, speaks of dispersing agency across the human *and* nonhuman worlds in networks of actants, thereby democratizing it in a radical sense. Ingold (2000) has sought to recast the human agent as already inescapably embedded in the world and not abstractable from it (see also Nash, 2005). I do not intend to adjudicate here on the ontological status of Latourian actants or on the persuasiveness of Heideggerian-sounding proposals about being-in-the-world. Philosophical interrogation is not my quarry. My concern is decidedly more modest. It is simply to allocate to landscape some role in setting limits on what observers can coherently say about it. I want to allow space for the thought that nature has some part to play in the theories that are constructed about it.

Such moves open up the possibility of reflecting in one way or another on the role of landscape in the generation and circulation of scientific knowledge claims. In this connection, it is suggestive to consider the ways in which natural historians who conducted research in Arctic landscapes responded to Darwin's theory—which was born of field observations in the temperate and tropical worlds. For Darwin the abundance and hyperfecundity of tropical nature was so overwhelming that he felt that describing it to an untraveled European was like trying to convey the experience of color to a blind man (Martins, 2000). Writing to a friend in August 1832 on the luxuriance of the Brazilian vegetation, he mused: "it was realizing the visions in the Arabian nights—The brilliancy of the Scenery throws one into a delirium of delight" (Darwin to Frederick Watkins, 18 August 1832, as quoted in Burkhardt & Smith, 1985, p. 260). As for the temperate world, it was his field experiments in the landscapes around Down House that furnished him with critical data. "My observations," he told the botanist Joseph Dalton Hooker in June 1857, "though on so infinitely a small scale, on the struggle for existence begin to make me a little clearer how the fight goes on"[1]. Recall, too, Wallace's (1878) observation that it was only in the equatorial latitudes that "a comparatively continuous and unchecked development of organic forms" had taken place and thus that in those regions "evolution has had a fair chance" (p. 122).

In the world of the Russian Arctic, things were different. Working in conditions where nature displayed no plenitude, no superabundance, no swarming life forms, the vocabulary of overpopulation and struggle between species just did not seem right. In that environment there developed a tradition of evolutionary zoology emphasizing cooperation in which the Malthusian components of natural selection were systematically expunged. As Peter Kropotkin later summarized the work of the St. Petersburg naturalists who carried out their inquiries in the Siberian wilderness and the Russian steppes, "*We see a great deal of mutual aid,* where Darwin and Wallace see *only struggle*" (as quoted in Todes, 1989, p. 104).

The Canadian north provides a useful comparator. Although there was a signal absence of response to Darwin among Canadian practitioners of geology, several botanists broached the subject during the early 1860s (Berger, 1983). The dominant motif in their endeavors was the fundamental significance of struggle against the vicissitudes of a harsh landscape. But, as in Russia, it was not struggle between species; instead it was struggle against an unyielding Precambrian shield. Success required inherited modification, and the idea of environmentally induced adaptation and acclimatization was resorted to so as to account for vegetational patterns. Such circumstances encouraged the agricultural reformer William McDougall to suggest in 1854 that the Lamarckian principle of the inheritance of acquired characteristics provided a viable explanation. Yet Darwinian language was embraced to a much greater degree in Canada than it was in Russia. George Lawson, for example, found J. D. Hooker's Darwinian account of arctic biogeography compelling, and George Dawson happily resorted to Darwinian vocabulary in his 1878 anthropological studies of the Haida people of Queen Charlotte Islands (Zeller, 1999).

Both landscapes, I venture to suggest, had some role to play in the production of the scientific theories that were constructed about them.

Political Ecology

Of course, the agency of landscape, as I have hinted, is neither monocausal nor unmediated. Rather, it is inflected in complex ways by what I call the political ecology of science, the view that nature is inescapably read through the lens of cultural politics and that the knowledge claims that manifest themselves in particular settings are the compound product of nature's agency *and* cultural hermeneutics. Latour (2004) has promoted the term 'political ecology' to argue against the conventional bifurcation between nature and culture and to urge their recasting into a new "collective" that "accumulates the old powers of nature and society in a single enclosure before it is differentiated once again into distinct powers" (p. 238). The ontological rearrangements that Latour envisages are intended to decompose nature as a specific sphere of reality—whose existence was always in any case a political constitution—and to reconceive of humans and nonhumans alike as members of "an assembly of beings *capable of speaking*" (Latour, 2004, p. 62). This simultaneous dissolution of "the social" and "the natural" and the surfacing of mediators through which forces act are, as I understand I, what Latour means by the "sociology of translation" (Latour, 2005, p. 106)—his preferred designation for what has become known as actor-network-theory (p. 106). These proposals clearly resonate with those who have been engaged in the task of complicating the assumed boundary lines between human and nonhuman. As Sarah Whatmore (2002) characterized it, as she inaugurated her project on "hybrid geographies," the "forays" that practitioners in the social sciences and humanities make "into the domain of natural sciences have swelled, so a plethora of 'things' has been trespassing into the company of the social[,] unsettling the conduct of its study" (p. 1).

The political ecology of science that I conceive is less metaphysically ambitious, less concerned with the epistemology of agency, less oriented toward taxonomic rearrangement than these proposals. Rather, it is intended to highlight the ways in which scientific knowledge of the natural world is politically constituted in different ways in different settings. In the Russian case that I have just reviewed, for example, the particular construction of Darwinism that crystallized there was at least in part the product of a landscape hermeneutic shaped in dialogue with Malthusian demographics. Both on the political left and right in Russia, Malthus's atomistic conception of society had already been castigated, mostly since the 1840s, as a cold, soulless, and mechanistic product of English political economy. Malthus may have rationalized poverty and inequity in England, but his commentators were certain that his theory would not apply in a harmonious Russia. It ran foul of Russian visions of a cohesive society that would jeopardize the cherished peasant commune (Todes, 1989). Indeed, according to Shaw and Oldfield (2007), the very understanding of what landscape is was construed in Russia in very particular ways reflecting political and ideological preoccupations.

In Canada the relative lack of response to Darwinism during the 1860s, 1870s, and 1880s sprang in part from an ingrained Baconianism that prioritized collecting and classification over theoretical speculation. The absence of public controversy as indifference turned into advocacy had much to do with the ways in which religious leaders found it possible to incorporate evolutionary thinking into their progressivist conceptions of historical change. In addition, its later adoption was bound up with romantic nationalist notions of "the north as a source of liberty, physical strength, and hardiness of spirit" (Zeller, 1999, p. 99). Darwin delivered a scientific framework that could feed the nation's resolve to overcome its harsh environment and mold a race fitted for survival. As Zeller (1998) puts it, "inhabitants of northern lands somehow acquired the mental and physical hardiness that destined them to thrive there. Biogeographical theories anthropomorphized northern forms ... that successfully 'invaded' southerly lands and moved in as 'denizens'" (p. 27).

The political ecologies of science manifest themselves in other ways, too, not least through the role that cultural politics play in shaping scientific knowledge and its circulation. Let me illustrate. In the decades around 1800, scientific speculations about human origins were massively freighted with political cargo. The debate between the Scottish jurist Lord Kames and the American moral philosopher Samuel Stanhope Smith is illustrative. Kames (1774) brought the full weight of his scholarship to bear on the question of the role climate played in racial differentiation. In his view, Montesquieu's resort to climate as the explanation for human variation was simply mistaken. As he read the record of the human past, he readily came to the conclusion that climate did not make human varieties; rather, human varieties were made for different climates. The implications—hesitant though he seemingly was to adopt them—were plain: human diversity was primitive, not derived, and different races were fitted for particular places. Such a conclusion was deeply troubling to Smith, whose *Essay on the Causes of the Variety of Complexion and Figure in the Human Species*, which first appeared in 1787 and then in an expanded form in 1810, had Kames in its crosshairs from the start. To Smith, polygenism was obnoxious scientifically, religiously, and—most critically—politically. Why? He spelled it out clearly in the 1810 edition of the work: "that the denial of the unity of the human species tends to impair, if not entirely destroy, the foundations of duty and morals, and, in a word, of the whole science of human nature. No general principles of conduct, or religion, or even of civil policy could be derived from natures originally and essentially different from one another" (Smith, 1810/1965, p. 149). In a setting where the new nation had overturned monarchy, inherited privilege, and established religion as the grounds of civic authority, universal human nature was the only foundation on which an orderly polity could be erected. In the early days of the new American Republic, a confidence in a common human constitution was precisely the philosophy that was needed if public virtue was not to dissipate. The realities of American geopolitics thus profoundly informed Smith's response to Kames's palaeoanthropological proposals (see Livingstone, 2008).

Print Culture

In the wake of the great flurry of interest in the history of the book, evidenced particularly in Johns's monumental *The Nature of the Book* (1998), there has been a growing recognition that there is a spatiality as well as temporality to textual productions of all kinds. The task of "bringing geography to book," to use James Ryan's neat phrasing, is opening up a host of critical questions revolving around print culture and the production, consumption, and circulation of knowledge (Ryan, 2003). As I have pointed out elsewhere (Livingstone, 2005), a suite of multidimensional geographies potentially manifest themselves, including the charting of the material spaces of book production, the distributional networks of mass print, the cultural topography of book buying, and the social morphology of lending libraries. And, of course, print culture extends far beyond the history and geography of the book. As Ogborn (2007) shows, many different modes of writing were implicated in the making and maintaining of the English East India Company. Heraldic manuscripts, political pamphlets, stock listings, official regulations, and many more were implicated in the construction of economic, governmental, and trading knowledges (Ogborn, 2007). Scientific enterprises are no less characterized by textual mulitiplicity; conventional published findings take their place alongside what Jardine (2000) calls "routinely authored works—instrument handbooks, instruction manuals, observatory and laboratory protocols," all of which are basic to the regulation of empirical practices (p. 401; see also Topham, 2000).

My own interest in print culture has tended to revolve around the located nature of hermeneutics and what I call the geographies of reading (Livingstone, 2003b). The basic thought is that texts are differently encountered in different settings and, thus, as they travel their meaning is transformed by the venues in which readers find themselves. Consider first an example from the sphere in which the practice of hermeneutics originally emerged, the interpretation of religious texts. To a great many slaveholders in the antebellum American South, a plain, unadorned reading of the Bible seemed to sanction the slave system, and there was no need to turn to secular science or unorthodox readings of Genesis to support it. To them, such perfidious projects would only defile a laudable religiously sanctioned institution and weaken the foundations of southern patriarchal communalism. Scriptural arguments in support of slavery abounded (Genovese & Fox-Genovese, 2005; Stout, 2006). The clergyman George Armstrong pronounced in *The Christian Doctrine of Slavery* (1857) that abolitionism had sprouted from the infidel breed of philosophy that had inspired the French Revolution. The theology professor George Howe, in an unrestrained attack on the polygenist lectures of Josiah Nott, repudiated efforts to justify slavery in the language of biology, preferring instead the vocabulary of the Bible, under which ancient and modern slaveholders "lived, protected and unrebuked" (Howe, 1850, p. 487). Similar cases could readily be elaborated.

In the northern states, by contrast, many readers interrogated the Bible for its abolitionist possibilities. Some ferreted out passages mandating the release of slaves after a set number of years; others pointed out that proslavery theologies conveniently ignored practices like polygamy, which enjoyed Old Testament sanction every bit as much as slavery. Still others dwelt less on specifics than on making a generalized moral case against slavery's inhumanity.

What became clear was that there simply was no such thing as a politically neutral, straightforward reading of the text. Hermeneutics just *were* shaped by the cultural conditions and political stance of commentators. As Holifield (2003) argues, the slave issue precipitated a move away from what he calls a Baconian hermeneutic by introducing a historical consciousness that insisted on the need to locate biblical texts in the time and place of their writing (pp. 494–504). The American Civil War—as Noll (2006) perceptively suggests—precipitated, and was precipitated by, a hermeneutic crisis.

The reading of scientific texts, theories, and reputations is no less susceptible to located interpretations, as recent research has amply disclosed. Let me illustrate. Much of the inspiration for this work has come from Secord's (2001) analysis of the reading of the early Victorian "sensation," Chambers's (1844) anonymously published *Vestiges of the Natural History of Creation*. What emerges from this remarkable work is that the meaning of the text was shaped in profound ways by where readers were located. In Liverpool, for example, the text's potential for inspiring urban reform was seized upon by the Mechanics' Institution with its anticlerical leadership, whereas in the Anglican-dominated space of the City's Literary and Philosophical Society it was met with alarm. In other urban settings other microgeographies of reading surfaced (Secord, 2001).

Rupke's (2005) "metabiography" of Alexander von Humboldt demonstrates how the reputation of even a single scientific figure could be differently forged in different venues. At various points in German political history from late-Prussian times through the Empire Period, the Weimar Republic, the Third Reich, and the divided Germany of post-1949, Rupke shows how the identity of Humboldt was recreated to suit the political sensibilities of the moment. Hence, he pauses to consider Humboldt the liberal democrat, Humboldt the Weimar *Kultur* chauvinist, Humboldt the Aryan supremacist, Humboldt the antislavery radical, and Humboldt the pioneer of globalization. All these projections press Rupke to the conclusion that the reading of an author's reputation is a located enterprise. Humboldt "has become a man with several lives," Rupke writes, "products of appropriation on behalf of geographically separate and chronologically successive socio-political cultures" (p. 16).

I myself have sought to map out something of the geography of reading Darwin. The meaning of Darwinism, I contend, was locally constituted in very different ways in different places. South Carolina naturalists read its monogenist anthropology as subversive of the racial basis of old southern culture, whereas many New Zealanders welcomed it because of its potential to underwrite the runaway triumphs of a settler society happily wiping out the Maori. In Scotland the high profile public controversy over the biblical criticism unleashed by William Robertson Smith (1875) cast any threat from Darwin in the shade, and few oppositional readings are evident. In Ireland Darwin's *On the Origin of Species by Means of Natural Selection; or, The Preservation of Favoured Races in the Struggle for Life* (1859) was differently read by Catholics and Protestants, by reviewers in Dublin, Belfast, and Derry. The meaning of Darwin's text was differently construed, depending on attitudes to the controversial Belfast address of John Tyndall at the 1874 meeting of the British Association for the Advancement of Science, to Catholic stipulations about the university curriculum, and to what was thought to be the implications of Darwin's theory for the management of population. In every case what Darwinism *was*, was locally constituted (Livingstone, 2005).

Yet more recently, Keighren (2006) has sought to trace out a book geography through his examination of the reception of Semple's *Influences of Geographic Environment* (1911). While disclosing something of its differential reading in different sites, Keighren also attends to the conditions that shaped encounters with this key text—notably, how it was read in the early twentieth century for the ways in which it could deliver a scientific methodology for a subject seeking disciplinary identity and institutional esteem. There are hints, too, that how Semple's book was read was shaped by reactions readers had to hearing her speak. What is also clear is that, although Semple herself shunned the term "geographic determinant" and claimed to speak "with extreme caution of geographic control" (Keighren, 2006, p. 530), she was routinely cast as an environmental determinist. Whatever she intended to communicate, readers persistently took her to be saying something else. As Fischer (2003) has tellingly remarked, "The wonder in reading is that the writer is never in control" (p. 344).

All of these instances serve to highlight the instability of scientific meaning and to demonstrate that, although texts may be immutable mobiles (though that proposal is itself doubtful), their meanings are entirely mutable. No single uncontested meaning can be distilled from them. The implication is that scholars seeking to come to terms with the spatiality of scientific knowledge must engage more intensively with the geographies of print culture than they have in the past. This reorientation will also involve taking seriously *literal* translation as well as the *metaphorical* translation of meaning that goes on across "contact zones" of one sort or another. As Elshakry (2008) points out in an analysis of the cultural politics of late nineteenth-century scientific translations into Arabic, the whole notion of knowledge in motion is rendered yet more problematic when translation and transliteration are involved. As she puts it, "the specific problem of finding appropriate and meaningful lexical equivalents in cross-lingual scientific discourse" (p. 702) is a dilemma of very considerable proportions "for understanding the geography of knowledge" (p. 703) because it involves "questions of linguistic tradition, cultural purity and modernity itself" (p. 702) all played out "against the background of colonial rule and its resentments" (p. 702). Translations into Arabic of the very term *science* were profoundly implicated in the politics of language because they abutted on matters of "cultural authority, social change, and literary tradition" (p. 705).[2] When one realizes that there were no immediate Arabic equivalents for terms like *race, species,* and *evolution*, the dilemmas multiply alarmingly for translating a work like Darwin's *Origin of Species* and dramatically bring to the fore what Clifford (1988) pertinently refers to as the "politics of neologism" (p. 175).

Speech Space

I want finally to turn to a fourth theme that has, in my view, been only patchily developed in accounts of the geographies of scientific knowledge—spaces of speech. My own interest here lies in the connections between what I call location and locution, that is, on the ways in which settings both enable and constrain spoken communication (Livingstone, 2007). But there are other routes into this inviting zone. The remarkable range of sites of scientific conversation is itself worthy of scrutiny. Alongside learned societies like the Royal Society or the laboratories that gentlemen naturalists had constructed in their own homes, the new scientific conversation of the seventeenth and eighteenth centuries spilled out into public houses and drawing rooms, coffeehouses, and parlors (Fara, 2004; Terrall, 1996; Walters, 1997). In the kitchen Joseph Addison observed during a visit to one particular household, he found women discoursing on the usefulness of mathematical learning (Meyer, 1995, p. 24). For the twentieth century, a good deal of work has focused on the discourse of lab technicians in an effort to show how science is made in particular settings (Latour & Woolgar, 1979; Traweek, 1988).

As for the Victorian period, Secord (2007) has begun the task of depicting the rich array of venues in which scientific conversation took place. He calls to mind the fundamental importance of speech in an oral culture where verbal presentation took precedence over scientific print, and of the rich array of sites where scientific conversations could take place—at high table, bedsides, scientific societies, dining clubs, soirees, and tea rooms. He provides rich description of what might be called conversation management as hosts consulted etiquette manuals for suitable topics for polite discussion and ways to prevent fashionable table talk from degenerating into vulgar shoptalk. Some subjects, like botany, were "in"; others, like mathematics or phrenology, were "out." Indeed, even in elite scientific spaces, Secord insists that verbal communication "remained central to the presentation of new scientific work" (p. 30).

In this vein, I want to dwell on the ways in which social spaces both shape, and are shaped by, speech. What can and cannot be said in particular venues, how things are said, and the way they are heard are all implicated in the production of knowledge spaces. In different arenas there are protocols for speech management; there are subjects that are trendy and subjects that are taboo. In public spaces and in camera, in formal gatherings and in private salons, in conferences and consultations, in courtrooms and churches, in clinics and clubs—in all these venues different things are speakable (and unspeakable) about scientific claims. In every case the setting sets limits on what can be spoken; the social space conditions what is heard. And individuals moving between these spaces adjust their speech—code-switching I believe it is called—to suit the setting. In so doing, as Burke (2004) points out, they are "performing different 'acts of identity' according to the situation in which they find themselves" (p. 6). In other words, the control of speech space is intimately connected with the maintenance of identity. Spaces of speech, of course, are also spaces of silence. There are always voices that are absent, or are not allowed to speak, or are denied access. In colonial societies, as Scott (1985) powerfully reminds his readers, the oppressed can rarely let their voices be heard. No doubt for different reasons, but with not dissimilar effects, those people marginalized in scientific debates find their voices unwelcome in science's privileged sites.

Let me just touch on two kinds of speech space and the ways in which their elucidation might illuminate the geographies of scientific circulation. First, family space. A sensitivity about what could be spoken at home was something about which leading scientists sometimes reflected with their close associates. Thus J. D. Hooker mused in a letter to Charles Darwin in which, as Brooke (2007) puts it, "the social pressures for conformity were perfectly explicit" (p. 16):

> It is all very well for Wallace to wonder at scientific men being afraid of saying what they think . . . . Had he as many kind and good relations as I have, who would be grieved and pained to hear me say what I think, and had he children who would be placed in predicaments most detrimental to children's minds . . . he would not wonder so much. (J. D. Hooker to C. Darwin, October 6, 1865, as quoted in Burkhardt, 2002, pp. 261–265)

Such concerns easily spilled over into anxieties about how a wider public might set constraints on what one was comfortable saying. Darwin certainly felt such moral pressure. He found it "a fearfully difficult moral problem about the speaking out on religion" (Brooke, 1985, p. 40). This circumstance, of course, makes it extraordinarily difficult to ascertain just precisely what he *did* think about certain

subjects. For, as Day (2008) has observed, the "penchant for deliberate and sometimes dissembling cultural self-fashioning could be particularly conspicuous when
religion was the subject of conversation" (p. 55).

In yet more public institutional spaces, how scientific claims were talked about
required care. Alexander Winchell, who lost his chair at Vanderbilt University
over evolution, had mused in an explanatory letter to the readers of the Nashville
American on June 15, 1878: "I have always taken pains, in my lectures at Nashville,
to avoid the utterance of opinions which I supposed were disapproved of by the officers of the University" (as quoted in Alberstadt, 1994, p. 108).[3] He evidently did
not succeed in his tongue tactics. In Belfast, in the aftermath of the assault that the
religious establishment had received from Tyndall's taunting speech at the British
Association meeting of 1874, pulpits, platforms, and presbytery meetings talked
about little else. The series of evening lectures that they organized in the city that
winter marked out the hermeneutic horizon against which the theory was judged for
more than a generation. The speech space that Tyndall helped crystallize set boundaries on what could be said and heard about Darwinian evolution, and the local
almanac for 1875 railed against the "very bad taste" that Thomas Henry Huxley and
John Tyndall exhibited in their recent addresses to the British Association. They
had infringed oral propriety. As MacIlwaine (1874–1875) was at pains to point out
at the Belfast Naturalists' Field Club at its winter session that year, talk of religious
belief in a scientific setting was "a violation of the rules of good taste"; Tyndall's
"reckless" incursion into theology and metaphysics at the meeting of the British
Association was thus nothing short of "reprehensible" (p. 82). In Boston, the students of Louis Agassiz could talk about the new Darwinian theory only in secrecy
for fear of their teacher's ire. Years later, Nathaniel Shaler reflected that to be caught
in such conversations "was as it is for the faithful to be detected in a careful study of
heresy" (as quoted in Livingstone, 1987, p. 28). Such concerns might help explain
why, when many of Agassiz's students did become evolutionists, they turned to the
Neo-Lamarckian version, which retained notions of inherent progress, rather than
to the orthodox Darwinian model.

Indeed, high profile clashes, like the infamous altercation between T. H. Huxley
and Samuel Wilberforce, Bishop of Oxford, at the 1860 meeting of the British
Association for the Advancement of Science, cannot be understood, in my view,
without attending to whether or not decorum was breached during the row. Matters
of etiquette and good taste were certainly in the minds of some observers who
reflected on the occasion. Frederic William Farrar, theological writer and later
Canon of Westminster recalled that what the bishop said was neither vulgar nor inso-
lent, but flippant, particularly when he seemed to degrade the fair sex by pondering
whether anyone—whatever they thought about their *grandfather*—would be willing
to trace their descent from an ape through their *grandmother*. In Farrar's opinion,
everyone recognized that the bishop "had forgotten to behave like a gentleman" and
that Huxley "had got a victory in the respect of *manners* and *good breeding*" (as
quoted in Lucas, 1979, p. 327). And yet, although later writers placed Huxley on
the side of good breeding, at the time both the *Athenaeum* and *Jackson's Oxford
Journal* thought him discourteous (Lucas, 1979). The boundaries of civility shifted

over the decades. As White (2003, p. 65) puts it, Huxley's frankness "still seemed unruly and discreditable" in 1860, whereas Wilberforce's ally Richard Owen, who had "once seemed honest and polite, appeared disreputable and ill-mannered" by later standards. So ... did the bulldog bite the bishop, or did the bishop badmouth the bulldog (see Livingstone, 2009)? It all depends on the character of the speech space that summer afternoon.

It does not take much imagination to make the transfer to today. Whether the conversation is about genetically modified crops, global warming, stem cell research, intelligent design, the commercial use of bio-organs, cold fusion, laser-guided weapon systems, or even the social construction of scientific knowledge, interlocutors are usually well aware of the immediate speech space they are occupying and what would constitute a violation of its rhetorical decorum. All these are controversial subjects, of course, but it is not difficult to entertain the thought that partisans for some particular scientific perspective are only too happy to police conversational arenas to outlaw rival theories. To ascertain just what role speech spaces continue to play in the circulation of knowledge seems to me to be a promising line of inquiry.

Conclusion

Scientific knowledge is a geographical phenomenon. It is acquired in specific sites; it circulates from location to location; it transforms the world. As students of the spatiality of science have pursued their inquiries, conventional distinctions between the natural and the social, the local and the global, and the scientific and the political have been rendered more and more troublesome. At the same time, attention to the role of material objects like specimens and samples that trace out their own dynamic geographies as they move around the world is opening up new and fertile lines of investigation. In this chapter I have sought to further supplement the agenda for geographical studies of scientific knowledge and practice by calling attention to the role of landscape in knowledge enterprises, to the political ecology of science, to the critical significance of print culture in the circulation of scientific claims, and to the place—and places—of speech in scientific culture. My reason for doing so is that science shapes and is shaped by the physical world; science produces and is produced by cultural politics; science generates and is generated by textual encounters; science is made and remade by how it is talked about. Landscape agency, political ecology, print culture, and speech space, I contend, are fundamental to the ongoing task of illuminating the geographies of scientific knowledge.

Notes

1. Retrieved August 10, 2009 from http://www.darwinproject.ac.uk/darwinletters/calendar/entry-2101.html.
2. I am most grateful to the author for sharing her pre-published analysis with me.
3. Winchell's letter is reproduced *in extenso* in Alberstadt (1994).

References

Alberstadt, L. (1994). Alexander Winchell's preadamites—A case for dismissal from the Vanderbilt University. *Earth Sciences History, 13*, 97–112.

Armstrong, G. (1857). *The Christian doctrine of slavery.* New York: Charles Scribner's.

Berger, C. (1983). *Science, God, and nature in Victorian Canada.* Toronto: University of Toronto Press.

Brooke, J. H. (1985). The relations between Darwin's science and his religion. In J. Durant (Ed.), *Darwinism and divinity: Essays on evolution and religious belief* (pp. 40–75). Oxford, England: Basil Blackwell.

Brooke, J. H. (2007). Response. *Historically Speaking, 8*(7), 16–17.

Browne, J. (1996). Biogeography and empire. In N. Jardine, J. A. Secord, & E. C. Spary (Eds.), *Cultures of natural history* (pp. 305–321). Cambridge, England: Cambridge University Press.

Burke, P. (2004). *Languages and communities in early modern Europe.* Cambridge, England: Cambridge University Press.

Burkhardt, F. H. (Ed.). (2002). *The correspondence of Charles Darwin* (Vol. 13). Cambridge, England: Cambridge University Press.

Burkhardt, F. H., & Smith, S. (Eds.). (1985). *The correspondence of Charles Darwin* (Vol. 1). Cambridge, England: Cambridge University Press.

Camerini, J. (1996). Wallace in the field. *Osiris, 11*, 44–65.

Chambers, R. (1844). *Vestiges of the natural history of creation.* London: Churchill.

Cittadino, E. (1990). *Nature as the laboratory: Darwinian plant ecology in the German empire, 1880–1900.* Cambridge, England: Cambridge University Press.

Clifford, J. (1988). *The predicament of culture.* Cambridge, England: Cambridge University Press.

Collins, H. M. (1988). Public experiments and displays of virtuosity: The core-set revisited. *Social Studies of Science, 18*, 725–748.

Cooper, A. (2006). Houses and households. In D. C. Lindberg & R. L. Numbers (Series Eds.) & K. Park & L. Daston (Eds.), *The Cambridge history of science: Vol. 3. Early modern science* (pp. 224–237). Cambridge, England: Cambridge University Press.

Darwin, C. (1859). *On the origin of species by means of natural selection; or, the preservation of favoured races in the struggle for life.* London: John Murray.

Day, M. (2008). Godless savages and superstitious dogs: Charles Darwin, imperial ethnography and the problem of human uniqueness. *Journal of the History of Ideas, 69*, 49–70.

Dorn, H. (1991). *The geography of science.* Baltimore: Johns Hopkins University Press.

Dritsas, L. (2005). From Lake Nyassa to Philadelphia: A geography of the Zambesi expedition, 1858–6. *British Journal for the History of Science, 38*, 35–52.

Eamon, W. (2006). Markets, piazzas, and villages. In D. C. Lindberg & R. L. Numbers (Series Eds.) & K. Park & L. Daston (Eds.), *The Cambridge history of science: Vol. 3. Early modern science* (pp. 206–223). Cambridge, England: Cambridge University Press.

Elshakry, M. (2008). Knowledge in motion: The cultural politics of modern science translations in Arabic. *Isis, 99*, 701–730.

Fara, P. (2004). *Pandora's breeches: Omen, science and power in the Enlightenment.* London: Pimlico.

Finnegan, D. (2008). The spatial turn: Geographical approaches in the history of science. *Journal of the History of Biology, 41*, 369–388.

Fischer, S. R. (2003). *A history of reading.* London: Reaktion Books.

Galison, P., & Thompson, E. (Eds.). (1999). *The architecture of science.* Cambridge, MA: MIT Press.

Genovese, E. D., & Fox-Genovese, E. (2005). *The mind of the master class: History and faith in the southern slaveholders' worldview.* New York: Cambridge University Press.

Gieryn, T. F. (2002). Three truth-spots. *Journal of the History of the Behavioral Sciences, 38*, 113–132.

Gieryn, T. F. (2006). The city as truth-spot: Laboratories and field-sites in urban studies. *Social Studies of Science, 36*, 5–38.

Hill, J. (2006a). Travelling objects: The wellcome collection in Los Angeles, London and beyond. *Cultural Geographies, 13*, 340–366.

Hill, J. (2006b). Globe-trotting medicine chests: Tracing geographies of collecting and pharmaceuticals. *Social and Cultural Geography, 7*, 365–384.

Holifield, E. B. (2003). *Theology in America: Christian thought from the age of the puritans to the civil war.* New Haven: Yale University Press.

Howe, G. (1850). Nott's lectures. *Southern Presbyterian Review, 3*, 426–490.

Ingold, T. (2000). *The perception of the environment: Essays in livelihood, dwelling and skill.* New York: Routledge.

Jardine, N. (2000). Books, texts, and the making of knowledge. In M. Frasca-Spada & N. Jardine (Eds.), *Books and the sciences in history* (pp. 393–407). Cambridge, England: Cambridge University Press.

Johns, A. (1998). *The nature of the book: Print and knowledge in the making.* Chicago: University of Chicago Press.

Johns, A. (2006). Coffeehouses and print shops. In D. C. Lindberg & R. L. Numbers (Series Eds.) & K. Park & L. Daston (Eds.), *The Cambridge history of science: Vol. 3. Early modern science* (pp. 320–340). Cambridge, England: Cambridge University Press.

Kames, H. H., Lord (1774). *Sketches of the history of man* (Vol. 2). Edinburgh: W. Strahan and T. Cadell.

Keeling, A. M. (2007). Charting marine pollution science: Oceanography on Canada's Pacific coast, 1938–1970. *Journal of Historical Geography, 33*, 403–428.

Keighren, I. M. (2006). Bringing geography to the book: Charting the reception of *Influences of geographic environment. Transactions of the Institute of British Geographers, New Series, 31*, 525–540.

Kohler, R. E. (2002). *Landscapes and labscapes: Exploring the lab-field border in biology.* Chicago: University of Chicago Press.

Latour, B. (1999). *Pandora's hope: Essays on the reality of science studies.* Cambridge, MA: Harvard University Press.

Latour, B. (2004). *The politics of nature: How to bring the sciences into democracy* (Catherine Porter, Trans.). Cambridge, MA: Harvard University Press.

Latour, B. (2005). *Reassembling the social: An introduction to actor-network-theory.* Oxford, England: Oxford University Press.

Latour, B., & Woolgar, S. (1979). *Laboratory life: The social construction of scientific facts.* Beverly Hills, CA: Sage.

Lawrence, C., & Shapin, S. (Eds.). (1988). *Science incarnate: Historical embodiments of natural knowledge.* Chicago: University of Chicago Press.

Livingstone, D. N. (1987). *Nathaniel Southgate Shaler and the culture of American science.* Tuscaloosa, AL: University of Alabama Press.

Livingstone, D. N. (2003a). *Putting science in its place: Geographies of scientific knowledge.* Chicago: University of Chicago Press.

Livingstone, D. N. (2003b). Science, religion and the geography of reading: Sir William Whitla and the editorial staging of Isaac Newton's writings on biblical prophecy. *British Journal for the History of Science, 36*, 27–42.

Livingstone, D. N. (2005). Science, text and space: Thoughts on the geography of reading. *Transactions of the Institute of British Geographers, New Series, 30*, 391–401.

Livingstone, D. N. (2007). Science, speech and space: Scientific knowledge and the spaces of rhetoric. *History of the Human Sciences, 20*, 71–98.

Livingstone, D. N. (2008). *Adam's ancestors: Race, religion, and the politics of human origins.* Baltimore: Johns Hopkins University Press.

Livingstone, D. N. (2009). Myth 17. That Huxley defeated Wilberforce in their debate over evolution and religion. In R. L. Numbers (Ed.), *Galileo goes to jail and other myths about science and religion* (pp. 152–160). Cambridge, MA: Harvard University Press.

Lucas, J. R. (1979). Wilberforce and Huxley: A legendary encounter. *The Historical Journal, 22,* 313–330.

MacIlwaine, W. (1874–1875). Presidential address, *Proceedings of the Belfast Naturalists' Field Club,* 81–99.

Martins, L. (2000). A naturalist's vision of the tropics: Charles Darwin and the Brazilian landscape. *Singapore Journal of Tropical Geography, 21,* 19–33.

Meusburger, P. (2000). The spatial concentration of knowledge: Some theoretical considerations. *Erdkunde, 54,* 352–364.

Meyer, G. D. (1995). *The scientific lady in England, 1650–1760.* Berkeley: University of California Press.

Nash, L. (2005). The agency of nature or the nature of agency? *Environmental History, 10.* Available online at http://www.historycooperative.org/journals/eh/10.1/nash.html

Naylor, S. (2005). Introduction: Historical geographies of science—Places, contexts, cartographies. *British Journal for the History of Science, 38,* 1–12.

Noll, M. A. (2006). *The civil war as a theological crisis.* Chapel Hill: University of North Carolina Press.

Ogborn, M. (2007). *Indian ink: Script and print in the making of the English East India Company.* Chicago: University of Chicago Press.

Park, K., & Daston, L. (2006a). Introduction: The age of the new. In D. C. Lindberg & R. L. Numbers (Series Eds.) & K. Park & L. Daston (Vol. Eds.), *The Cambridge history of science: Vol. 3. Early modern science* (pp. 1–17). Cambridge, England: Cambridge University Press.

Park, K., & Daston, L. (Vol. Eds.). (2006b). In D. C. Lindberg & R. L. Numbers (Series Eds.), *The Cambridge history of science: Vol. 3. Early modern science.* Cambridge, England: Cambridge University Press.

Parry, B. (2004). *Trading the genome: Investigating the commodification of bio-information.* New York: Columbia University Press.

Parry, B. (2006). New spaces of biological commodification: The dynamics of trade in genetic resources and 'bioinformation'. *Interdisciplinary Science Studies, 31,* 19–31.

Polanyi, M. (1959). *The study of man.* Chicago: University of Chicago Press.

Powell, R. C. (2007). Geographies of science: Histories, localities, practices, futures. *Progress in Human Geography, 31,* 309–329.

Reid-Henry, S. (in press). *The Cuban cure: Reason and resistance in global science.* Chicago: University of Chicago Press.

Rupke, N. A. (2005). *Alexander von Humboldt: A metabiography.* Frankfurt am Main: Peter Lang.

Rupke, N. A. (2009). Nature as culture: The example of animal behaviour and human morality. (Unpublished manuscript)

Ryan, J. (2003). History and philosophy of geography: Bringing geography to book, 2000–2001. *Progress in Human Geography, 27,* 195–202.

Schiebinger, L., & Swan C. (Eds.). (2005). *Colonial botany: Science, commerce, and politics in the early modern world.* Philadelphia: University of Philadelphia Press.

Scott, J. (1985). *Weapons of the weak: Everyday forms of peasant resistance.* New Haven: Yale University Press.

Secord, J. A. (2001). *Victorian sensation: The extraordinary publication, reception, and secret authorship of Vestiges of the Natural History of Creation.* Chicago: University of Chicago Press.

Secord, J. A. (2004). Knowledge in transit. *Isis, 95,* 654–672.

Secord, J. A. (2007). How scientific conversation became shop talk. In A. Fyfe & B. Lightman (Eds.), *Science in the marketplace* (pp. 23–59). Chicago: University of Chicago Press.

Semple, E. C. (1903). *American history and its geographic conditions on the basis of Ratzeöl's system of anthropo-geography.* Boston: Houghton Mifflin.

Semple, E. C. (1911). *Influences of geographic environment on the basis of Ratzel's system of anthropo-geography*. New York: Holt.

Shapin, S. (1998). Placing the view from nowhere: Historical and sociological problems in the location of science. *Transactions of the Institute of British Geographers, New Series, 23*, 5–12.

Shaw, D. J. B., & Oldfield, J. D. (2007). Landscape science: A Russian geographical tradition. *Annals of the Association of American Geographers, 97*, 111–126.

Smith, S. S. (1810/1965). *Essay on the causes of the variety of complexion and figure in the human species* Cambridge, MA: Harvard University Press. (Expanded ed.)(Original work published 1787)

Smith, W. R. (1875). Bible. *Encyclopaedia Britannica* (9th ed., Vol. 3, q.v.). Edinburgh: A. and C. Black.

Stout, H. S. (2006). *Upon the altar of the nation: A moral history of the American Civil War*. New York: Viking.

Terrall, M. (1996). Salon, academy, and boudoir: Generation and desire in Maupertuis's science of life. *Isis, 87*, 217–229.

Todes, D. P. (1989). *Darwin without Malthus: The struggle for existence in Russian evolutionary thought*. Oxford, England: Oxford University Press.

Topham, J. R. (2000). Scientific publishing and the reading of science in nineteenth-century Britain: An historiographical survey and guide to sources. *Studies in History and Philosophy of Science, 31A*, 559–612.

Traweek, S. (1988). *Beamtimes and lifetimes: The world of high energy physicists*. Cambridge, MA: Harvard University Press.

Tyndall, J. (1874). *Address delivered before the British Association assembled at Belfast, with addition*. London: Longmans, Green, and Co.

Wallace, A. R. (1878). *Natural selection and tropical nature*. London: Macmillan.

Walters, A. N. (1997). Conversation pieces: Science and politeness in eighteenth-century England. *History of Science, 35*, 121–154.

Whatmore, S. (2002). *Hybrid geographies: Natures cultures spaces*. London: Sage.

White, P. (2003). Thomas Huxley: Making the "man of science". Cambridge, England: Cambridge University Press.

Withers, C. W. J. (2002). The geography of scientific knowledge. In N. Rupke (Ed.), *Göttingen and the development of the natural sciences* (pp. 9–18). Göttingen: Wallstein Verlag.

Withers, C. W. J. (2007). *Placing the Enlightenment: Thinking geographically about the age of reason*. Chicago: University of Chicago Press.

Zeller, S. (1998). Classical codes: Biogeographical assessments of environment in Victorian Canada. *Journal of Historical Geography, 24*, 20–35.

Zeller, S. (1999). Environment, culture, and the reception of Darwin in Canada, 1859–1909. In R. L. Numbers & J. Stenhouse (Eds.), *Disseminating Darwinism: The role of place, race, religion, and gender* (pp. 91–122). Cambridge, England: Cambridge University Press.

Global Knowledge?

Nico Stehr

When Wells published his monumental two-volume *The Work, Wealth and Happiness of Mankind* in 1931, one of the memorable metaphors that appeared in the first volume was the "abolition of distance." Wells saw the analysis of the conquest of distance as the central reason for the urgent need to generate "sound common ideas about work and wealth," but his analysis of the economy of the modern world was also a treatise about what people now call the globalization process. Because of the abolition of distance, geography "has become something very different from what it was" (Wells, 1931, p. 4).[1]

Wells (1931) is fascinated by the success story of the rapidly developing communications technologies in particular. Their triumph leads him to anticipate that the "whole world will be a meeting place" (p. 170); more than that, he is convinced that "mankind seems to be approaching a phase when we shall realize and think almost as if we had one mind in common" (p. 160).[2] I take it, therefore, that Wells is an early proponent of the idea that global worlds of knowledge are possible, although his prediction is driven by the belief in the efficacy of technically displacing distance, an act that greatly facilitates the transmission of "facts." Nonetheless, the question of whether a "global" world of knowledge might emerge among modern societies is still largely unexplored from the viewpoint of the social sciences today.[3] I begin examining the notion of global knowledge by noting a few contemporary thoughts about the global context of my own observations.

First, the nature and value of "local" or "indigenous" knowledge is undergoing reevaluation against the backdrop of globalization processes. More specifically, assessment of the economic value of local knowledge, that is, the appraisal of indigenous knowledge as a source of sustainable development, is changing

N. Stehr (✉)
Karl Mannheim Chair for Cultural Studies, Zeppelin University, Friedrichshafen
D-88045, Germany
e-mail: nico.stehr@t-online.de

This chapter is a considerably expanded and revised version of an oral presentation at the third symposium on "Knowledge and Space" in Heidelberg, June 27–30, 2007. I am grateful to Paul Malone for his editorial assistance.

(Fernando, 2003). An initiative of the World Bank (1998) that proposes to utilize indigenous knowledge in developmental processes represents one side of this reevaluation. Much of its other side is represented by growing efforts to protect indigenous agrarian, technical, and botanical knowledge (see Agrawal, 2002; Leach & Fairhead, 2002; World Commission, 2004).

Second, various forms of discourse, including policy discussions, increasingly emphasize the rapid and perhaps unstoppable diffusion of scientific forms of knowledge and therefore also both the local and ubiquitous (global) relevance of that knowledge. Although experts with narrow competencies generate contemporary scientific knowledge in highly specialized settings like laboratories and field stations, such knowledge is expected to have ubiquitous qualities (Livingstone, 2003). When the World Bank (1999) refers to the social risks of the modern "information revolution," more particularly to "knowledge gaps" and "information problems" between the developing and the developed world, it presupposes that they are reducible, notably through institutional transformations in developing regions of the world. But if these knowledge differentials within and among societies can be overcome with great difficulty (if at all), then "new" powerful elites with considerable economic strength should emerge (Radhakrishnan, 2007).

Third, the presumed ease with which specialized knowledge travels is apparent in recent transnational treaties designed to protect intellectual property. (This point applies to information as well, given that people often use the words *knowledge* and *information* synonymously). One example is the World Trade Organization's Trade-related Aspects of Intellectual Property Rights (TRIPS) Agreement (Retrieved August 13, 2009, from http://www.wto.org/english/tratop_e/trips_e/trips_e.htm).

Science studies and the sociology of knowledge stress that the generation of knowledge takes place in distinct places and under special circumstances. The spaces and circumstances touch the substance of what is produced. This observation raises certain questions. Can such context-dependent knowledge travel? If it indeed moves around, does it travel as part of the baggage of knowledge carriers, such as scientists and engineers? If it is diffused, is it not transformed by transcending features of its origins? If it is transformed as the result of travel, is it really true that knowledge can become global only if the places of its origins achieve global presence and become globally known? Despite political and economic intentions to protect knowledge that otherwise threatens to become "homeless"—or, as H. G. Wells put it, effortlessly connected all over the world—the possibility or nature of the limits to globalizing knowledge is an open question.

The open issues can be formulated more succinctly. Is "placeless knowledge" a matter of the growing worldwide "standardization" of organizational or social forms (e.g., states, business enterprises, and scientific communities), or are there forms of knowledge (and its carriers) that increasingly transcend locality (see Freeman, 2005) and coactively promote converging social and knowledge contexts around the world?

Approach

I present observations on modern globalizing worlds of knowledge[4] in a series of steps, starting with an outline of a simple conceptual model that distinguishes between the horizontal and vertical diffusion of knowledge. After then briefly conveying my conceptual view of knowledge, I delimit my topic in greater detail and refer to allegedly preexisting global worlds of knowledge, indicating where they are supposed to be found. This skeptical enumeration of knowledge that has already achieved the status of global knowledge leads into a series of assumptions meant to aid me as explanatory guides. The observations thereafter are divided into two parts. The first is a discussion of those aspects raising hope for the fair chance that globalizing worlds of knowledge might exist. The second, longer, section draws attention to social processes and features of knowledge that make the implementation of global worlds of knowledge appear less than imminent. My observations conclude with a brief summary.

A Simple Conceptual Model

Assuming that a strictly uniform distribution of knowledge is impossible (see Luckmann, 1982), I want to distinguish between the existence and nonexistence of horizontal and vertical integration of worlds of knowledge across social boundaries. Horizontal and vertical integration of knowledge may be linked to three aspects. The first is the degree to which the social bases of producing and consuming knowledge are concentrated (i.e., embedded in persons, equipment, books, journals, laboratories, and the like). An increase in the horizontal convergence of the social bases of the production of knowledge could mean, for example, that the social distribution across the world of scholars producing knowledge is broader than it used to be. Or it could mean that the means for producing knowledge are simply less concentrated than they once were. Increase in the vertical penetration of knowledge—as opposed to concentration of knowledge as an instrument of domination in the hands of the powerful—could imply that the boundaries between highly specialized knowledge and common knowledge in different constituencies are no longer as imposing as in past periods of civilization. The second aspect to which the vertical and horizontal integration of knowledge may be linked is agenda-setting. It arises with the emergence of cross-boundary questions, problems, and issues that require advice and solutions. The third aspect is an increase in the similarity of the "means" in which knowledge is embedded and of the claims that are advanced, independently of political boundaries.[5] Knowledge may have horizontally penetrated much of the world, but the same need not be the case for vertical integration within and across social worlds. At this point the issue of the presence and resistance of indigenous knowledge comes into play.

Knowledge About Knowledge

I propose to define knowledge as the possibility of taking action, as a capacity (resource) for action.[6] My view of knowledge as a model for reality[7]—and not as model *of* reality—is that an open connection exists between action and knowledge.[8] This openness is as true of the production of knowledge as to the application of knowledge.[9] Knowledge is a resource available as practically needed. It gains distinction through its capacity to change reality.[10] Capacities to transform reality are contingent by nature; they are just as contingent as human life itself.[11] Knowledge in general, and scientific knowledge in particular, is not only a potential means of access to the secret of the world but also the coming into being of the world. Are such kinds of knowledge, however, the emergence of *one* world?

Delimitations

To delimit my reflections on the concept of global worlds of knowledge, I observe three restrictions. First, one can speak of global worlds of knowledge only if knowledge is *obligated* to travel (however such travel is accomplished), in other words, only if the producers and the consumers of knowledge are not identical. This separation also applies to knowledge anchored in social constructions or technological artifacts, that is, to knowledge that indeed travels as part of a product but whose content is not directly "consumed." An example is the knowledge inscribed in material substances (e.g., pharmaceuticals, foods, and technological artifacts such as cars) or in nonmaterial entities (e.g., images or symbols). Global worlds of knowledge relate to the "horizontal" distribution of knowledge, not its "vertical" distribution (e.g., according to class or social stratum).

Second, global knowledge as an accomplishment starts as "external" knowledge that travels to a social environment from which it was absent. The external knowledge in question must also have to be free of charge. Its disclosure has to be voluntary and accessible to all recipients.

Third, I do not concern myself with the worldwide dissemination of the institution of science or with the observation that educational systems, professional occupations, and universities exist in almost every society.

Global Worlds of Knowledge as Thought Experiments, Normative Frames and Business Plans

As far as I can tell, global or globalizing worlds of knowledge have hitherto been approached, and even achieved, primarily in normative speculations, by decree, as a thought experiment, or as a business plan. Consider the following mélange of such ideas:

1. *Freedom is the daughter of knowledge.* One of the first modern advocates of global worlds of knowledge and their social and political potential was the social philosopher Otto Neurath. As an émigré in 1941, Neurath, in his capacity as a consultant for urban renewal in the slums of an English industrial town, wanted to give himself the title of Consulting Sociologist of Human Happiness. The question of how to increase the happiness of all, according to Neurath, was the question of the conditions for democratizing knowledge. In almost all his writings (e.g., Neurath, 1991), he advocated the democratic right to global worlds of knowledge, at least insofar as it can be asserted within a society.

   The medium of this democratization was Neurath's system of speaking signs, the pictorial language he scientifically designed to instruct the broad masses objectively about their situation (see Hartmann & Bauer, 2002). Like many other fields of knowledge when they originate, this one, which came to be called cybernetics, was seen as the center of imperial knowledge, as a sanctuary of global worlds of knowledge in theory and in practice. (More recently, the environmental sciences, such as climatology, have similarly been imputed the capacity to generate knowledge claims whose policy consequences are global and globalizing.)

2. *Knowledge must be a global public asset* (e.g., Stiglitz, 1999, 2007, pp. 103–118; World Bank, 1998). From an economic viewpoint, this position means that knowledge does not have the characteristics otherwise typical for economic assets—rivalry and excludability. Though that lack may well be true for the day-to-day social stock of knowledge, it probably is not for additional (i.e., new) knowledge. Additional knowledge turns a profit. Even in the sciences, the privatization of knowledge is an increasingly prevalent phenomenon.

3. *Global worlds of knowledge exist as business plans and in thought experiments.* The Internet, which is globally accessible for the most part, teems with enterprises in whose business plans "global worlds of knowledge" have already been implemented and offered for sale. An example is Global Knowledge—an organization with 1,300 employees around the globe, a New York investment bank, and the business slogan "Experts teaching experts."[12] This firm, as its very name suggests, offers knowledge applicable worldwide for improving companies' earnings.

These normative visions, promising business plans, and pronouncements about global worlds of knowledge are often exposed as Eurocentric prejudices that deny nonwestern actors the ability to govern themselves successfully, to create notable cultural artifacts, or to produce enduring contributions to rational discourse. Such illuminating sociohistoric contextualizations demonstrate the disputability of claims to global relevance and substantiate the skepticism prompted nowadays by unduly expansive global statements (see Gough, 2002). In the following sections, I concentrate on the question of how and why it is possible or improbable that global worlds of knowledge come to exist.

Attributes of Knowledge that Appear to Promote the Chances of Its Global Dissemination

A list of the attributes of knowledge that promote the chances of global worlds of knowledge is relatively brief. I have in mind two characteristics that make it pointless to regulate knowledge and therefore limit its diffusion around the globe. The first is the understandable fact that a bias against the availability of resources on the commons exists in many cultures. Experience has shown that unrestricted access to resources on the commons leads to their overconsumption. To be sure, not all resources fall victim to this law, for some of them are apparently inexhaustible or the limits on their growth are either of a different nature or do not figure at all. Despite the debate about the end of science in some quarters of the philosophy of science, knowledge is obviously one such resource. The primary task is thus to promote access to the resource, not to try futilely protecting it from consumption.

The second of the two attributes promoting the chances of global worlds of knowledge is that knowledge realizes itself. There seem to be two parallel strands of argumentation leading to the conclusion that knowledge stubbornly realizes itself and, therefore, that any form of knowledge policy conceived to regulate or restrict the realization and dissemination of knowledge is doomed to failure from the beginning (Stehr, 2003, 2004). In the one strand, the utilization of knowledge is built into the structure of knowledge itself. The manufacture of knowledge implies its realization and prevents any control over the application of knowledge. I call to mind the once much-discussed conceptual definition of the various inherent epistemological interests of the sciences (Habermas, 1964). The often invisible category of technical epistemological interest refers to the fact that fabricated knowledge (or objects) of this sort entails an impetus to self-manifestation. More precisely, that kind of knowledge is connected from the outset to an abiding sense of utility.[13]

A related assumption about characteristics of knowledge that facilitate its dissemination arises from the difference between implicit and codified knowledge, which are said to be the two poles between which the development of knowledge ranges. Economic material constraints guarantee that this evolutionary course of knowledge development is ever more frequently observable in modern societies (see Cowan & Foray, 1997).[14]

The Constraints on Global Worlds of Knowledge

In this section I focus on two specific forms of constraints on the development of globalizing worlds of knowledge: (a) intra- and inter-societal limits (e.g., a society's legal practices) and (b) characteristics that can be traced directly back to certain characteristics of knowledge itself.

Max Weber was among the first social scientists to be concerned in general terms with the obstacles to the application of modern technology and modern scientific thought outside the Western world. He dealt in much less detail with the reasons for the use of technology in, say, revolutionizing the production process in Western societies. In the present context, though, the weight Weber placed on the special value that cultural practices have for social development and the migration of knowledge can help fill a gap in his comparative analysis. His perspective reveals, namely, that it was the carriers of scientific knowledge who played a decisive part in the social acceptance of updated technology and science. Indeed, it is generally necessary for the carriers of scientific and other technical knowledge to have a certain degree of autonomy from the ruling strata of society in order to be able to break with traditional forms of knowledge. In addition, they need access to an organizational infrastructure in order to disseminate their knowledge. Weber's perspective therefore makes it possible to underscore the role of cultural carriers and cultural obstacles that inhibit the dissemination of knowledge and technology.

Since World War II, the basic conditions governing the production of scientific and technical knowledge in the United States (and, consequently, in other nations) have undergone two radical changes. The first revolution, launched in the United States about 60 years ago by Vannevar Bush (1945) (see also O'Mara, 2005), decoupled the production of knowledge from politicomilitary goals. This segregation led to the expansion of the great research universities in the United States and can be seen as having improved conditions for the possible emergence of global worlds of knowledge. Knowledge was able to circulate more easily than it had before that point.

Stressing the autonomy of the sciences as a precondition for acquiring "uncontaminated" knowledge and thereby allowing for "scientific progress" and practical knowledge alike, such varied theorists of science as Karl Popper, Michael Polanyi, John Bernal, and Vannevar Bush were in agreement during this period. To these thinkers the utility of knowledge was therefore the daughter of truth, and truth was the supreme goal of scientific work.

The second transformation, which has acted to preserve the limits on the free circulation of knowledge and thereby lessen the likelihood of knowledge becoming a global public asset (Stiglitz, 1999), dates from the U.S. Senate's passage of the Bayh-Dole Amendment in 1980. This legislation permitted researchers to patent their discoveries even when they had been made with the help of public funding.[15] Before the amendment, few patents had been issued on government-funded research; the passage of the legislation was followed by a dramatic shift of research activities to the private sector (see Kennedy, 2002).[16]

The concomitant changes in the dissemination of knowledge, or the material substrata of knowledge, are becoming especially clear in the biomedical sciences. Certain material substrata of research, such as cell lines or certain mice whose "production" is cost intensive, are no longer freely obtainable by other groups of researchers. Similarly, the availability of certain commercially and academically produced methods, data, and research results has been eliminated by competition. In short, access to the conditions for generating new knowledge and acquiring the results of that knowledge production is being increasingly regulated (Pottage, 1998).[17]

To put the argument positively, trade in services and products is an important *inter*societal vehicle for the dissemination of knowledge and the development of globalizing worlds of knowledge. An expansion of worldwide trade, especially the dismantling of trade barriers for developing economies, could lead to an unintended worldwide diffusion of ideas and knowledge as easily as to the reduction of gaps in information and knowledge in the world. To be sure, one ought not imagine this process as simply the transmission of knowledge and artifacts but rather as the development of hybrid forms of knowledge.

The notion that knowledge protects itself has a supply side and a demand side (Kitch, 1980, pp. 711–715).[18] Either knowledge is extremely hard to steal or the great difficulty in profiting from it means that scarcely anyone is interested in stealing it. On the supply side, self-protecting knowledge means that the use of knowledge must be closely tied to the capacity to mobilize cognitive abilities that are both rare and difficult to articulate. The difficulties of using knowledge secondarily and of transporting it are a function of factors such as the manner in which knowledge is organized. The self-protection of knowledge also means that knowledge is anchored in a given knowledge infrastructure, such as the ability to learn how to learn, and thus can neither circulate freely nor be easily reconstituted.

On the demand side, the self-protecting qualities of knowledge might include processes associated with characteristics of knowledge or with its application. One example is the high depreciation of knowledge, the fact that acquired knowledge quickly loses its value relative to the costs of acquisition and future profits. And with certain forms of knowledge (e.g., a famous painting or a very rare book, such as the Gutenberg Bible), the attendant rights of ownership may be easily attributable by others and are therefore of value primarily to the owner. One can accelerate the attrition of knowledge and information by behaving according to that information. For example, following advice to buy a certain stock does not guarantee that the investment will subsequently appreciate. The high degree of wear and tear on information implies that "by the time someone steals the information, it is worthless, which in turn means there is no incentive to steal it" (Kitch, 1980, p. 714).

Prospects for the Future

Despite the age of economic globalization, there are as yet no global worlds of knowledge. Globalization does not erase the fundamental social law of the simultaneity of the nonsimultaneous. As the result of structural and cultural differences among societies, the likelihood of a global world of knowledge is still only a remote possibility.

Notes

1. His description of the forces and consequences of abolishing distances was in many ways quite modern: "Almost all our political and administrative boundaries, the 'layout' of the

human population, have become, we begin to realize, misfits. Our ways of doing business, dealing with property, employing other people, and working ourselves have undergone all sorts of deformation because of the 'change of scale' in human affairs. They are being altered under our eyes, and it behooves us to the very best of our ability to understand the alterations in progress" (Wells, 1931, p. 4).

2. To Wells, the information technologies of his day and those whose wide use he anticipated, such as television, represented benign social forces rather than devices that amplify political power. His view of these technologies contrasts with earlier and later scholarly and artistic images conveying the possible panoptic principles of the "information society" (Jeremy Bentham, George Orwell, Michel Foucault, and Gary Marx).

3. My discussion of global worlds of knowledge is restricted to modern contemporary societies and therefore takes note of the widely unquestioned view that the origin of *divided* worlds of knowledge is a modern phenomenon. As Durkheim (1955/1983) accentuated in his lectures on pragmatism: "It is in the very early ages that men, in every social group, all think in the same way. It is then that uniformity of thought can be found. The great differences only begin to appear with the very first Greek philosophers. The Middle Ages once again achieved the very type of the intellectual consensus. Then came the Reformation, and with it came heresies and schisms which were to continue to multiply until we eventually came to realize that everyone has the right to think as he wishes" (p. 76).

4. As explored in this chapter, the notion of "global" knowledge cannot possibly mean "common knowledge" (or a common mind) in the strict sense. Common knowledge of an event, for example, is implied in a group of agents "if each knows it, if each one knows that the other knows it, if each one knows that each one knows that the others know it, and so on" (Geanakoplos, 1992, p. 54). Global knowledge, if knowledge ever approaches such a state, will always be imperfect, asymmetric, moving, and stratified, but it will also have elements of commonality and convergence that distinguish it from "merely" local knowledge.

5. Costs may not be a major barrier to the access to knowledge. As Olson (1996) notes, it seems that "most advances in basic science can be of use to a poor country only after they have been combined with or embedded in some product or process that must be purchased from firms in the rich countries" (p. 7). But Koo's (1982) case study on South Korean economic advances from 1973 to 1979—research that attempts to quantify the costs of important technologies from abroad—shows that "the world's productive knowledge is, for the most part, available to poor countries, and even at relatively modest cost" (Olson, 1996, p. 8). During that period in Korea, the cost of acquiring knowledge from abroad (loosely defined) amounted to less than 1.5% of the *increase* in that country's GDP.

6. I do not mean that the capacity to act theoretically stands at the beginning and precedes action. The capacity to act is acquired by means of action and is realized by means of carrying action to its completion (see Janich & Weingarten, 2002, p. 115; Stehr, 1991).

7. The idea that knowledge is a model for reality that illuminates, discloses, and transforms, but also displaces, reality resonates with Borgmann's (1999) conception of the "recipe" (or plans, scores, and constitutions) as a "model of *information for reality*" (p. 1).

8. This view of knowledge is related to a series of parallel sociological observations, such as the concept of culture as a generalized capacity that helps actors produce strategies for action. As Swidler (1986) underlines, one must imagine culture as a "'tool kit' or repertoire from which actors select differing pieces for constructing lines of action . . . . People may have in readiness cultural capacities they rarely employ; people know more culture than they use" (p. 277).

9. Accordingly, I do not employ a performative conception of knowledge in my argument. A performative definition of knowledge would be consistent with "performative utterance" as described in Austin's (1962) linguistics—in other words, with the claim that knowledge *does* that which it describes (see also Osborne & Rose, 1999). It is at least worth considering, however, that the modern process of knowledge production is increasingly concerned with expanding the creation of performative knowledge (see Stehr, 2003, pp. 107–108) and,

hence, with exploring a progressive interpenetration of "theory and practice." The discovery of a gene, for example, is simultaneously the test for that gene.

10. However, the particular reputation that knowledge has won in society through its ability to facilitate the alteration of social and natural processes also has an increasingly prominent drawback in the political confrontation with newly discovered knowledge. This liability is the fact that the capacity of knowledge to affect socially anchored classifications or boundaries of feasibility leads to efforts to discipline newly discovered knowledge (see Stehr, 2003).

11. But the thesis that human knowledge is contingent (see Easton, 1991, 1997, pp. 39–40) does not preclude convergence of knowledge, say, in the sense of multiple, but independent, discoveries of possibilities for action.

12. Information retrieved August 22, 2007, from http://www.globalknowledge.com/training/generic.asp?pageid=2&translation=English

13. With technical artifacts there is also a widespread assumption that technical developments are inherently deterministic, that is, that they are marked from the start by a deep-seated destiny excluding ambiguous or even alternative forms of development and thus permitting no "interpretative flexibility" (Pinch & Bijker, 1984, pp. 419–424).

14. Cowan and Foray (1997) define the codification of knowledge as "the process of conversion of knowledge into messages which can be processed as information" (p. 596). These authors assume that information can be disseminated far more easily than noncodified knowledge, or at least at less marginal cost.

15. Assessment of the Bayh-Dole Amendment of 1980 varies considerably. Some observers lament the fact that the size of the "knowledge commons" has shrunk through increased patenting and that corporate interests have intruded on university campuses and laboratories. Supporters of the amendment prefer to call it a huge success, for it has increased the technology transfer from campuses and laboratories for the purpose of developing new products and services (Kennedy, 2005).

16. On the patenting of *living* entities in the United States, see Kevles (2001). Pharmaceutical products in particular appear to profit from being patented, especially because even minimal alteration of a medication's composition can result in a new patent. Broadening the definition of knowledge to include "branding" or the reputation of products (e.g., those of Coca-Cola, Nike, or Mercedes) greatly limits the dissemination—in the sense of "imitation"—of the knowledge they carry (see Kay, 1999, p. 13).

17. Though conceding that the protection of intellectual property by means of patents, copyright laws, and comparable state-sanctioned norms results in short-term monopolistic profits, conventional, utilitarian economic theorists emphasize that it also provides incentives to innovation that ultimately serve the common good (Arrow, 1962, pp. 616–617; Bentham, 1839, p. 71; Pigou, 1924, pp. 151; Smith, 1776/1976, pp. 277–278). By contrast, Boldrin and Levine (2002) state that the competitive market is very likely to be in the position to reward entrepreneurial investments in research and development. In their opinion, patent laws are therefore superfluous and, regarding the prices of new products, for example, even harmful (Hirshleifer, 1971; Plant, 1934a, 1934b).

18. The proposition that modern knowledge may have self-protecting characteristics is not primarily concerned with certain inherent features that make knowledge akin to a private asset. (The self-protecting attributes of knowledge may have existed particularly in earlier centuries, when scientific knowledge was protected from laymen by being formulated in one of the least accessible languages and was thus, so to speak, automatically protected). Rather, it refers to context-dependent institutional attributes that hinder a simple dissemination of knowledge. One of them in modern society is the lack of access to the educational system and its intellectual capital.

References

Agrawal, A. (2002). Indigenous knowledge and the politics of classification. *International Social Science Journal, 59*, 287–297.

Arrow, K. J. (1962). Economic welfare and the allocation of resources for invention. In R. R. Nelson (Ed.), *The rate and direction of inventive activity: Economic and social factors* (pp. 609–625). Princeton, NJ: Princeton University Press.

Austin, J. L. (1962). *How to do things with words*. London: Oxford University Press.

Bentham, J. (1839). *A manual of political economy*. New York: G. P. Putnam.

Boldrin, M., & Levine, D. K. (2002). *Perfectly competitive innovation*. Federal Reserve Bank of Minneapolis, Research Department, Staff Report 303. Retrieved March 8, 2008, from http://www.dklevine.com/papers/pci23.pdf

Borgmann, A. (1999). *Holding on to reality. The nature of information at the turn of the millennium*. Chicago: University of Chicago Press.

Bush, V. (1945). As we may think. *Atlantic Monthly, 176*(1), 101–108.

Cowan, R., & Foray, D. (1997). The economics of codification and the diffusion of knowledge. *Industrial and Corporate Change, 6*, 595–622.

Durkheim, E. (1983). *Pragmatism and sociology* Cambridge: Cambridge University Press. (J. Whitehouse, trans.)(Original work published 1955)

Easton, D. (1991). The division, integration, and transfer of knowledge. In D. Easton & C. S. Schelling (Eds.), *Divided knowledge: Across disciplines, across cultures* (pp. 7–36). Newbury Park, CA: Sage.

Easton, D. (1997). The future of the postbehavioral phase in political science. In K. R. Monroe (Ed.), *Contemporary empirical political theory* (pp. 13–46). Berkeley: University of California Press.

Fernando, J. L. (2003). NGOs and the production of indigenous knowledge under the condition of postmodernity. *Annals of the American Academy of Political and Social Science, 590*, 54–72.

Freeman, R. B. (2005). *Does globalization of the scientific/engineering workforce threaten US economic leadership?* (NBER Working Paper No. W11457). Available at SSRN: http://ssrn.com/abstract=755693

Geanakoplos, J. (1992). Common knowledge. *Journal of Economic Perspectives, 6*, 53–82.

Gough, N. (2002). Thinking/acting locally/globally: Western science and environmental education in a global knowledge economy. *International Journal of Science Education, 24*, 1217–1237.

Habermas, J. (1964). Dogmatismus, Vernunft und Entscheidung—Zur Theorie und Praxis in der wissenschaftlichen Zivilisation [Dogmatism, reason, and decision-making—On theory and practice in scientific civilization]. In J. Habermas (Ed.), *Theorie und Praxis* (pp. 231–257). Neuwied: Luchterhand.

Hartmann, F., & Bauer, E. K. (2002). *Bildersprache. Otto Neurath Visualisierungen* [Metaphorical language: Otto Neurath Visualizations]. Vienna: WUV.

Hirshleifer, J. (1971). The private and social value of information and the reward to innovative activity. *American Economic Review, Papers and Proceedings, 63*, 31–39.

Janich, P., & Weingarten, M. (2002). Verantwortung ohne Verständnis? Wie die Ethikdebatte zur Gentechnik von deren Wissenschaftstheorie abhängt [Responsibility without understanding? How the ethics debate on genetic engineering depends on its scientific theory]. *Journal for General Philosophy of Science, 33*, 85–120.

Kay, J. (1999). Money from knowledge. *Science and Public Affairs,* April, 12–13.

Kennedy, D. (2002). On science at the crossroads. *Daedalus, 131*, 122–126.

Kennedy, D. (2005). Bayh-Dole: Almost 25. *Science, 307*, 1375.

Kevles, D. J. (2001). *Patenting life: A historical overview of law, interests, and ethics*. Paper presented to the Legal Theory Workshop, Yale University, December 20, 2001.

Kitch, E. W. (1980). The law and the economics of rights in valuable information. *Journal of Legal Studies, 9*, 683–723.

Koo, B.-Y. (1982). *New forms of foreign direct investment in Korea*. Korean Development Institute (Working Paper No. 82–02).

Leach, M., & Fairhead, J. (2002). 'Manners of contestation': 'Citizen science' and 'indigenous knowledge' in West Africa and the Caribbean. *International Social Science Journal, 59*, 299–311.

Livingstone, D. N. (2003). *Putting science in its place: Geographies of scientific knowledge*. Chicago: University of Chicago Press.

Luckmann, T. (1982). Individual action and social knowledge. In M. von Cranach & R. Harré (Eds.), *The analysis of action: Recent theoretical and empirical advances* (pp. 247–265). Cambridge: Cambridge University Press.

Neurath, O. (1991). *Schriften. Band 3: Gesammelte bildpädagogische Schriften*. Vienna: Hölder-Pichler-Tempsky.

Olson, M., Jr. (1996). Big bills left on the sidewalk: Why some nations are rich, and others are poor. *The Journal of Economic Perspectives, 10*, 3–24.

O'Mara, M. P. (2005). *Cities of knowledge: Cold war science and the search for the next Silicon Valley*. Princeton, NJ: Princeton University Press.

Osborne, T., & Rose, N. (1999). Do the social sciences create phenomena? The example of public opinion research. *British Journal of Sociology, 50*, 367–396.

Pigou, A. C. (1924). *The economics of welfare*. London: Macmillan.

Pinch & Bijker. (1984). The social construction of facts and artefacts: Or how the sociology of science and the sociology of technology might benefit each other. *Social Studies of Science, 14*, 399–441.

Plant, A. (1934a). The economic aspects of copyright in books. *Economica, 1*, 167–195.

Plant, A. (1934b). The economic theory concerning patents for inventions. *Economica, 1*, 30–51.

Pottage, A. (1998). The inscription of life in law: Genes, patents, and bio-politics. *The Modern Law Review, 61*, 740–765.

Radhakrishnan, S. (2007). Rethinking knowledge for development: Transnational knowledge professionals and the 'new' India. *Theory and Society, 36*, 141–159.

Smith, A. (1976). *The wealth of nations*. Oxford: Oxford University Press. (Original work published 1776)

Stehr, N. (1991). *Praktische Erkenntnis* [Practical knowledge]. Frankfurt am Main: Suhrkamp.

Stehr, N. (2003). *Wissenspolitik* Die Überwachung des Wissens [Knowledge Politics: Policing knowledge]. Frankfurt am Main: Suhrkamp.

Stehr, N. (Ed.) (2004). *The governance of knowledge*. New Brunswick, NJ: Transaction Books.

Stiglitz, J. E. (1999). Knowledge as a global public good. In I. Kaul, I. Grunberg, & M. A. Stern (Eds.), *Global public goods* (pp. 308–325). Oxford: Oxford University Press.

Stiglitz, J. E. (2007). *Making globalization work: The next steps to global justice*. London: Allen Lane.

Swidler, A. (1986). Culture as action: Symbols and strategies. *American Sociological Review, 51*, 273–286.

Wells, H. G. (1931). *The work, wealth and happiness of mankind* (Vol. 1). Garden City, NY: Doubleday, Doran & Company.

World Bank. (1998). *Indigenous knowledge for development*. Initiative led by the World Bank in partnership with CIRAN/NUFFIC, CISDA, ECA, IDRC, SANGONet, UNDP, UNESCO, WHO, WIPO. Retrieved December 11, 2008, from http://www.worldbank.org/html/aft/IK

World Bank. (1999). *Knowledge for development*. World Development Report. New York: Oxford University Press.

World Commission on the Social Dimensions of Globalization. (2004). *A fair globalization: Creating opportunities for all*. Geneva: International Labour Office.

Part II
Mobilities and Centers

A Geohistorical Study of "The Rise of Modern Science": Mapping Scientific Practice Through Urban Networks, 1500–1900

Peter J. Taylor, Michael Hoyler, and David M. Evans

According to Wallerstein (1974, 2004b), the modern world-system emerged in the "long sixteenth century" (c. 1450–1650) as a European-based world-economy straddling the Atlantic. Its basic structure encompassed a division of labor that defined core and periphery zones of economic activity. During the long period of its establishment, the core zone moved from Mediterranean Europe to northwest Europe, reflecting the reorientation of Europe to the rest of the world. It is the processes that create and recreate the core zone that have generated the social changes that have ultimately led to the elimination of all alternative world-systems; by about 1900 the modern world-system was effectively global in scope. One of these core processes has been what is conventionally known as the "rise of modern science."

As well as core-periphery spatial structures, world-systems analysis recognizes structures of knowledge that have also changed in the unfolding of the modern world-system (Wallerstein, 1999, 2004a, 2004b). The process that is identified as the "rise of modern science" links these two structures.[1] In other words, adopting a world-systems analysis approach enables us to integrate the stories of changing science and changing geography through the modern world-system. The relevance of "science" to social needs emanating from the endless accumulation of capital,

M. Hoyler (✉)
Department of Geography, Loughborough University, Loughborough LE11 3TU, UK
e-mail: m.hoyler@lboro.ac.uk

P.J. Taylor
Department of Geography, Loughborough University, Loughborough LE11 3TU, UK
e-mail: p.j.taylor@lboro.ac.uk

D.M. Evans
Department of Geography, Loughborough University, Loughborough LE11 3TU, UK
e-mail: d.m.evans@lboro.ac.uk

With kind permission from Springer Science+Business Media: Minerva, A Geohistorical Study of "The Rise of Modern Science": Mapping Scientific Practice Through Urban Networks, 1500–1900, 46, 2008, 391–410, Peter J. Taylor, Michael Hoyler, and David M. Evans, © Springer Science+Business Media B.V. 2008.

the *differentia specifica* of the modern world-system, has encompassed both materialist reproduction (the work of science underpinning technology) and ideological reproduction (the idea of science underpinning progress).[2] This has culminated in science dominating modern structures of knowledge through two successful challenges to "traditional knowledges": first, a secular challenge to religious authority inherited from medieval knowledge structures; and second, an empirical challenge to the critical philosophical knowledge that replaced the sacred. By about 1900 specialized scientific disciplines had been created in recognizably modern universities (Ben-David & Zloczower, 1962; Wallerstein, 2004a). This chapter brings together these two parts of world-systems thinking, core-periphery and knowledge structures, by mapping the production of modern structures of knowledge through a detailed study of the changing geography of scientific practice in the modern world-system to 1900. We show it is a process located in the core of this world-system, but its geography is always more complex than specification as simply core process indicates. Sometimes scientific practice lags behind other core processes, at other times it is at the forefront of changes in core location.[3] We draw upon the encyclopedic research of Gascoigne (1984, 1987, 1992) for historical data to depict these geographies of science.

In his historical demographic study of "modern science," Gascoigne (1992) uses biographical data on 12,000 persons who practiced "science" from the early thirteenth century through to 1900.[4] The pattern of change he presents matches Wallerstein's chronology of core zone changes: There is no growth of scientific activity through the medieval period, but from sometime in the late-fifteenth century onwards "science" grows exponentially (Gascoigne, 1992, p. 550). Furthermore, during the "long sixteenth century," Italy initially grows to dominate in scientific activity but then stagnates to be "overtaken" by England, France, and Germany, and specifically Holland (Gascoigne, 1992, pp. 556–559; Ben-David, 1971).[5] Subsequently within northern Europe, it is Germany in particular that contributes most to the spectacular growth of "modern science" in the later years of Gascoigne's study. In this chapter we break down Gascoigne's national geographical categories into the actual urban places in which the science is practiced.[6] Further, we map specific urban networks of science by focusing on the "career" paths of one thousand "scientists" that Gascoigne has identified as "the most important in the period 1450–1900" (1992, p. 548; Gascoigne, 1987). From this source we are able to recreate some of the spatial dynamics that were integral to the practice of science as it grew to become and be central to modernity.

The argument proceeds through three sections. First, we set out the parameters of our study within the context of the rise of historical geographies of science. We argue that our contribution is to bring in particular social theories that augment our understanding of the rise of science as a world-systems process. Second, we introduce the data we use, explain how we analyze it, and discuss different levels at which our results may be interpreted. Third, we present our findings on the spatial dynamics of the "career" paths of leading scientists from 1500 to 1900. This highlights the geographies of connections across Europe in "the rise of modern science."

Historical Geographies of Science

Livingstone (2005) has identified a geographical turn in science studies and this has been particularly the case for historical studies of science. In a recent review, Naylor (2005) specifies three such historical geographies of science: the microgeographies of science, the places where scientific activity occurs; the broader contexts in which science exists, defined by scale—these are city, region, national and international contexts; and cartographies of science, the geographies within science discourses. Despite a conclusion that identifies a need to go "beyond place and culture" (p. 11), this framework is a very place-orientated conceptualization of geographies.[7] He does note that there are "many other ways of thinking geographically about the history of science" including thinking beyond "fixity" to "movements and circulations that help sustain . . . science," but the latter is conspicuous by its absence from the substance of his review.[8] This is a classic example of the tendency of much of the more humanistic approach in human geography to focus on place at the expense of flows.[9] In this chapter we take a more social science approach to our geography and bring flows to center stage.

A key understanding to derive from more humanistic approaches to the historical study of science is the diversity of roles that practitioners of science have played. Shapin (2006) emphasizes this heterogeneity by showing how the "man of science" has been cleric, government official, clerk, family tutor, domestic servant, gentleman, medical practitioner, as well as university scholar. The meanings of the science being practiced obviously varied with the roles being played. Thus, according to Harris (2006, p. 346), it is "anachronistic to speak in terms either of a 'scientific community' as a coherent group or of 'scientist' as a professional designation during the early modern period." There may have been no "scientific community" as Gascoigne purports to measure, but this does not mean there were not collective practices of knowledge production.[10] Shapin's "man of science" was also a "man of letters." Correspondence was very important: One of the Royal Society's foundation committees was the Correspondence Committee (Lux & Cook, 1998). It is this communicative aspect of scientific practitioners, the "Republic of Letters," that Harris (2006, pp. 347–348) focuses upon. Lux and Cook (1998) take a similar approach and highlight the problem of focusing on places with small groups, which they term "closed circles." They argue for more attention to be given to "open networks": The "assumption that all practices are local practices [is] undercutting the sense of a European-wide movement" in early modern scientific practice (p. 201). Thus, while respecting heterogeneity among scientific practitioners, it is important not to throw out the communicative baby with the community bathwater in studying the lineage of modern science.

A key advantage of employing a social science approach to the historical study of science is that it gives access to pertinent social theory. For instance, Harris (2006, p. 354), in his discussion of the Republic of Letters, draws upon Habermas's (1989) ideas to understand the overlapping and expanding networks of knowledge in early modern scientific practice as part of the new fabric of urban life that was the establishment of a public sphere. Lux and Cook (1998) provide a particularly creative adaptation of a social theory to historical circumstances. They use Granovetter's (1973) social network model, which salvages the specific importance of weak ties in diffusions of ideas. The argument is that strong ties tend to be inward-looking (closed) circles that play little part in passing ideas on. In contrast, weak ties are vital as efficient transmitters of ideas; being in between the closed circles, they form the vital links in the circulation of ideas. Lux and Cook (1998) show how the replacement of a more informal Paris institution by the new secretive Academy of Sciences in 1666 led to a loss of weak ties through which Paris had linked the southern half of scientific Europe to the north. Properly applied, these studies show social theory can furnish useful tools for the historical study of science.

We employ different areas of social theory from those reported above. As indicated at the start, our study is framed by Wallerstein's world-systems analysis.[11] We are concerned to describe and analyze one important core-making process: science practice. Our focus is on the creation of social spaces in the core zone of the modern world-system by agents (scientific practitioners) carrying out their (scientific) activities. Such activity reproduces spaces that sustain (scientific) trust between participants. In a general model of the production of social spaces, Castells (1996) identifies two types of social space: spaces of places and spaces of flows.[12] In the former type, contiguity facilitates face-to-face interaction, creating places of activity that generate trust. A market place is a classic example of such space; a laboratory is an equivalent scientific space. With spaces of flows trust is based upon indirect contact, distant communication through which trust is built up. Banking networks are a classic case of this form of space (you let your salary be paid to strangers); publication in a peer-reviewed journal is an equivalent example of a space of flows that sustains scientific practice. Of course, these two forms of social spaces do not exist in isolation; they need one another. Places are constituted by flows (input, throughput, output); flows are organized through places (nodes).[13] Therefore, the research choice is not which space to study but rather which space to use as the starting point of analysis.[14] Choosing where to start prioritizes one type of space over the other but does not necessarily neglect the space not chosen first. It is in this spirit that we use spaces of flows as the initial social space below. This is to focus on the dynamics before addressing the fixity: to recover historical networks and then to consider nodes in the network.

Whereas spaces of places are observable as relatively neat maps, the study of spaces of flows is far messier: The myriad overlapping networks, chains, circuits, and paths have been likened to dealing with a blizzard (Thrift, 1999, p. 272). The Republic of Letters referred to above can be interpreted as part of just such a space of flows. In this study we research the "career" paths of a population of scientists and aggregate them to show networks of workplace connections. This is an embodied space of flows that will have been a determinant of parts of the virtual Republic of Letters. The key advantage of dealing with aggregated "career" paths is that it provides a manageable universe of flows to delineate the geohistorical patterns in the "rise of modern science" as a material space of flows.[15]

Social flows are articulated through nodes, in this case the workplaces of scientists. Materialist urban theory that treats towns and cities as places of work can be brought into play at this point. We use the urban theory of Jacobs (1969) to make some sense of what happens to urban settlements when scientists (a university) come(s) to town. Her argument is that a city is a place within overlapping networks associated with unique internal complexity: A highly diverse urban place is a successful dynamic city. Scientific activity may be important in this process because, in Naylor's (2005) terms, as well as "science" creating "spaces and places for its own activities" (p. 3), it also "spatializes the world in a wide variety of ways" (p. 3). In other words, scientific activity can have influences beyond its own sphere. We use Jacobs's theory of city complexity to provide a new take on the perennial town-versus-gown conflict over place.

Data Construction and Analysis

The data for this study come from Gascoigne's (1987) chronology of the history of science. Part 2 of Gascoigne's work, entitled "The Social Dimension," provides "career" sketches of one thousand scientists, selected from an earlier, more comprehensive list (Gascoigne, 1984). Inclusion in the "top thousand" list was based on "the degree of importance accorded to each [scientist] in various biographical dictionaries and encyclopaedias and in histories of the individual sciences" (Gascoigne, 1987, p. ix).[16] Although inevitably based on some subjective judgment, a comparison of statistical data derived from this source with Gascoigne's (1984) original list and data compiled from the *Dictionary of Scientific Biography* shows a remarkable degree of consistency between the different sources (Gascoigne, 1992).

Biographical entries on individual scientists are arranged by country or region (Italy, France, Britain, Germany, Holland, Scandinavia, Switzerland, Eastern Europe, Russia, and the United States). Within each territory, individual "careers" are placed chronologically by decade of "career" start. Entries vary in length but generally list workplaces by town or city. Often, the workplaces refer to universities and other teaching institutions, but other sites of scientific engagement (e.g., courts, museums, botanical gardens, and observatories) are also listed. For data collection and analysis, we divided the available data into four centuries of major developments in the history of "modern science" (Fig. 1). In practical terms, this temporal division ensures a large enough number of scientists in each period to permit an analysis of key shifts by workplace rather than nationality of the scientists. More important, although centuries are arbitrary time periods, in this case they do relate to different phases in the "rise of modern science" as indicated by the labels attached to them in Fig. 1. Subsequent spatial analyses confirm the utility of this time frame.[17]

We initially recorded the "career" paths of listed scientists in a matrix that arrays scientists (columns) against workplaces (rows). For each century, one such matrix was created: 116 scientists × 61 places in the sixteenth century; 151 scientists × 85 places in the seventeenth century; 145 scientists × 88 places in the eighteenth century; and 422 scientists × 114 places in the ninenteenth century. Each cell of each

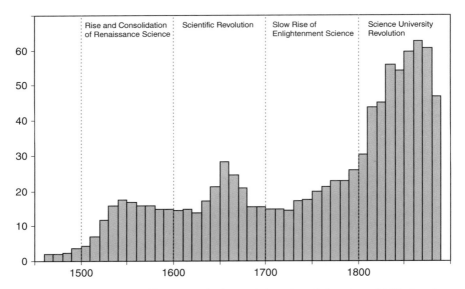

Fig. 1 Growth of the "scientific community," amended version of Gascoigne (1992), based on Gascoigne (1987). The number of "leading scientists" is given by decade of the "career" start (three-decade moving average)

matrix records presence or absence of a scientist in a particular location during that person's "career." The chronology of "career" stops is indicated by the sequence of allocated numbers (i.e., 1 for first workplace, 2 for second workplace, etc.).[18] In total, information was collected for 834 scientists for whom details of workplaces were available in the source.[19] This basic information allows a crude reconstruction of individual "career" paths. For the purpose of this study, we then converted each of the scientist × place matrices into a place × place matrix (ranging in size from 61 × 61 places in the sixteenth century to 114 × 114 places in the nineteenth century). Each cell in these inter-workplace matrices records, for the respective century, the number of "career" stops of leading scientists that link two specific locations.

For each of the four analyses two sets of results are presented below. First, for each place we count the number of scientists who spent part of their "careers" at that place. This tabulation provides a simple measure of the importance of a place; we refer to it as the nodality of the place within the overall network. In this way the most important scientific workplaces can be identified for each century. Second, we focus on the dyads (pairs of places) and count the number of scientists whose "career" paths encompassed both locations. In this way we can find the most important links between places. It is the latter that can be used to delineate the space of flows for each century.

Finally, before we present our results it is necessary to say something about interpretation. The findings below can be considered at two levels. First, at a literal level, they show actual patterns of "career" path characteristics over four centuries. This is interesting in its own right, but "career" paths in and of themselves are not our prime concern.[20] Second, at an inferential level, the results represent a general patterning of the "rise of modern science." Our conjecture is that our data and analysis indicates more than "career" paths; they are surrogates for much broader processes. Whether this is substantiated depends on the robustness of the data, model, and theory underpinning this extensive research. These have each been dealt with above, but the proof is in the pudding so to speak. We know of no other mapping study that attempts to recreate a detailed urban geography of the "rise of modern science," but we think our results are in line with what is known about this changing geography.

Geohistories of Scientific Career Paths

The basic geohistory of "modern science" can be gleaned from the workplaces of leading scientists: Fig. 2 shows the distribution of places where Gascoigne's leading scientists spent parts of their "careers" over four centuries.[21] For each map places are shown in proportion to the place with the most "career" stops. The patterns are not surprising but do reinforce our preconceptions of what was happening across Europe in terms of scientific practice. Clearly, in the sixteenth century northern and central Italy dominate in a "primate" pattern centered on Padua.[22] Beyond this core region there is a string of places just north and west of the Alps plus Paris, the English Cambridge–London–Oxford triangle, and a small German scattering. In the seventeenth century the pattern is very different: Padua continues to dominate in Italy, but across the Alps there are four other important centers at London, Leiden, Paris, and Jena. They constitute the only strong polycentric distribution we have uncovered in this research.[23] By the eighteenth century, Italian places have declined as scientific centers, and Paris now dominates in a weakly polycentric pattern. Finally, in the nineteenth century there is a reversion to a primate pattern, now centered on Berlin, with Germany in general dominating places of science.

Some of the statistics from which these maps were drawn are given in Tables 1, 2, 3, and 4, which list the leading places. In Tables 1, 2, and 3 (sixteenth, seventeenth, and eighteenth centuries, respectively), the lists of top places include those where at least five leading scientists spent part of their "career." Three pieces of information are given for each place: its scientist count, its proportion of the highest count, and its percentage of all scientist "career" stops for the century.

In Table 1 the primacy of Padua is clear, with that city having more than twice as many scientific "career" stops as Montpellier, which is ranked second. Furthermore, Padua has almost 14% of all 322 "career" stops we have recorded in the sixteenth century—by far the highest percentage we report for any century. The polycentricity of the seventeenth century is equally clear from Table 2, with three places covered by a range of just two stops and another four places with over half the "career"

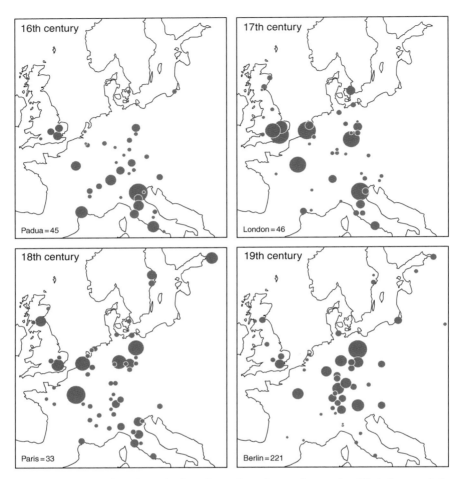

Fig. 2 Workplaces of leading scientists, sixteenth to nineteenth centuries. Workplace symbols proportional to place with most "career" stops in each century; perceptual scaling of symbols. Absolute numbers for top places are given in Tables 1, 2, 3, and 4

stops of the highest ranked place. Note that the percentages for overall stops, of which there were 472, are all below 10%. In the eighteenth century (Table 3) the growth of scientific "career" stops stalls, with only 346 being recorded. Paris has the highest percentage, with three other places showing over half of Paris's relatively low total. Note that Berlin, ranking second, is the highest ranked German place thus far. In this case the low level of polycentricity is further shown by the fact that the percentages of total stops are the lowest recorded in this study. Finally, in Table 4 equivalent results for the nineteenth century are shown, but with a total of 2,032 "career" stops recorded, the cut-off point for listing a place is set at 30. This change of criterion reflects a transformation of scale in the practice of

Table 1 Scientists' movements through places in the sixteenth century

Place	Absolute number	Proportion of highest	Percentage of total
Padua	45	1.000	13.98
Montpellier	22	0.489	6.83
Rome	20	0.444	6.21
Bologna	18	0.400	5.59
Basel	18	0.400	5.59
Paris	15	0.333	4.66
Pisa	14	0.311	4.35
London	12	0.267	3.73
Wittenberg	11	0.244	3.42
Tübingen	11	0.244	3.42
Ferrara	11	0.244	3.42
Cambridge	11	0.244	3.42
Nuremberg	10	0.222	3.11
Oxford	8	0.178	2.48
Vienna	7	0.156	2.17
Lyons	6	0.133	1.86
Geneva	6	0.133	1.86
Jena	5	0.111	1.55

Table 2 Scientists' movements through places in the seventeenth century

Place	Absolute number	Proportion of highest	Percentage of total
London	46	1.000	9.75
Leiden	44	0.957	9.32
Padua	44	0.957	9.32
Jena	38	0.826	8.05
Paris	38	0.826	8.05
Oxford	33	0.717	6.99
Cambridge	26	0.565	5.51
Bologna	14	0.304	2.97
Wittenberg	14	0.304	2.97
Copenhagen	13	0.283	2.75
Rome	13	0.283	2.75
Amsterdam	12	0.261	2.54
Basel	11	0.239	2.33
Montpellier	8	0.174	1.69
Venice	8	0.174	1.69
Leipzig	6	0.130	1.27
Pisa	6	0.130	1.27
St. Andrews	6	0.130	1.27
Aberdeen	5	0.109	1.06
Florence	5	0.109	1.06
Hamburg	5	0.109	1.06
Rostock	5	0.109	1.06

Table 3 Scientists' movements through places in the eighteenth century

Place	Absolute number	Proportion of highest	Percentage of total
Paris	33	1.000	9.54
Berlin	24	0.727	6.94
Leiden	22	0.667	6.36
Göttingen	21	0.636	6.07
London	16	0.485	4.62
St. Petersburg	15	0.455	4.34
Edinburgh	13	0.394	3.76
Halle	11	0.333	3.18
Padua	10	0.303	2.89
Uppsala	10	0.303	2.89
Bologna	9	0.273	2.60
Freiburg	8	0.242	2.31
Pisa	8	0.242	2.31
Pavia	7	0.212	2.02
Tübingen	6	0.182	1.73
Montpellier	5	0.152	1.45
Rome	5	0.152	1.45
Stockholm	5	0.152	1.45

Table 4 Scientists' movements through places in the nineteenth century

Place	Absolute number	Proportion of highest	Percentage of total
Berlin	221	1.000	10.88
Munich	108	0.489	5.31
Göttingen	99	0.448	4.87
Heidelberg	97	0.439	4.77
Leipzig	94	0.425	4.63
Paris	93	0.421	4.58
Würzburg	92	0.416	4.53
Bonn	84	0.380	4.13
London	68	0.308	3.35
Zurich	59	0.267	2.90
Strasbourg	52	0.235	2.56
Königsberg	48	0.217	2.36
Vienna	45	0.204	2.21
Tübingen	43	0.195	2.12
Breslau	41	0.186	2.02
Giessen	41	0.186	2.02
Halle	40	0.181	1.97
Jena	39	0.176	1.92
Marburg	39	0.176	1.92
Edinburgh	36	0.163	1.77
Freiburg	36	0.163	1.77
Cambridge	33	0.149	1.62
Erlangen	33	0.149	1.62
St. Petersburg	32	0.145	1.57
Kiel	31	0.140	1.53

science in Europe in the nineteenth century. The table confirms Berlin's primacy and Germany's dominance: Twenty of the places listed are German-speaking universities. Berlin also records a percentage of total "career" stops second only to Padua in the sixteenth century. However, it is somewhat short of the latter's percentage, reflecting a far broader network of university science places in the nineteenth century.

The previous results show where scientific practices were taking place across the four centuries but do not show actual connections, dyad links between places. We have constructed a set of four maps that illustrate the changing space of flows through which "modern science" has been constructed. The dyads we map are aggregates of "career" links between places. In Figs. 3, 4, and 5 (sixteenth, seventeenth, and eighteenth centuries, respectively), all dyads with at least two links in scientists' "career" paths are shown. In Fig. 3 sixteenth-century European science is shown unmistakably as a network of links centered on Padua. The strongest links are found in Italy and in the English triangle, but the latter is relatively isolated—Montpellier is clearly the second node of the network. In Fig. 4 there is a clear expansion of the network, not just geographically in its reach across the Alps but

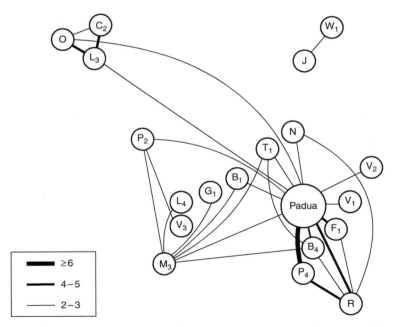

Fig. 3 Sixteenth-century networks of scientific practice. City codes: B_1 Basel, B_4 Bologna, C_2 Cambridge, F_1 Ferrara, G_1 Geneva, J Jena, L_3 London, L_4 Lyons, M_3 Montpellier, N Nuremberg, O Oxford, P_2 Paris, P_4 Pisa, R Rome, T_1 Tübingen, V_1 Venice, V_2 Vienna, V_3 Vienne, W_1 Wittenberg

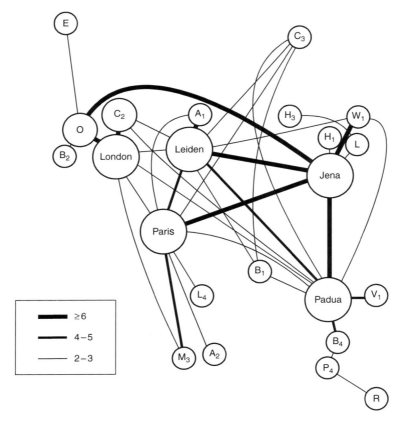

Fig. 4 Seventeenth-century networks of scientific practice. City codes: A_1 Amsterdam, A_2 Avignon, B_1 Basel, B_2 Bath, B_4 Bologna, C_2 Cambridge, C_3 Copenhagen, E Edinburgh, H_1 Halle, H_3 Helmstedt, L_2 Leipzig, L_4 Lyons, M_3 Montpellier, O Oxford, P_4 Pisa, R Rome, V_1 Venice, W_1 Wittenberg

structurally in the strength of its polycentric cohesiveness. Note that this diagram differentiates the main five centers listed in Table 3. Although London is ranked first in terms of "career" stops (Table 2), its position in the network is not so marked: The other four leading places all have more links than London, with Padua continuing to have the most links overall and Jena having by far the highest number of strong links. London's high number of "career" stops (Table 2) is based simply on unusually close links within the English triangle during the scientific revolution. Figure 5 shows the eighteenth-century pattern, which suggests a dissolution of the previous century's space of flows. The network appears to be breaking up into four subnets: northern Italy just surviving, a mainly Paris-centered French net, a London-centered British net (including Scottish Enlightenment places and Leiden), and a more dispersed German net. There are only three dyads linking the subnets:

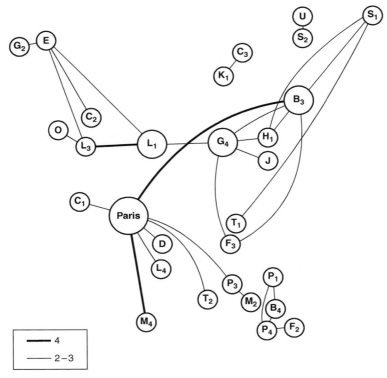

Fig. 5 Eighteenth-century networks of scientific practice. City codes: B_3 Berlin, B_4 Bologna, C_1 Caen, C_2 Cambridge, C_3 Copenhagen, D Dijon, E Edinburgh, F_2 Florence, F_3 Freiburg, G_3 Glasgow, G_4 Göttingen, H_1 Halle, J Jena, K_1 Kiel, L_1 Leiden, L_3 London, L_4 Lyons, M_2 Modena, M_3 Montpellier, O Oxford, P_1 Padua, P_3 Pavia, P_4 Pisa, S_1 St Petersburg, S_2 Stockholm, T_1 Tübingen, T_2 Turin, U Uppsala

Paris–Berlin, Paris–Pavia, and Leiden–Göttingen. This picture looks very much like the end of a process.

Because of the scale of change, the diagram for the nineteenth century cannot use the same dyad-size categories: In Fig. 6 all the dyads shown are in categories that were combined in the largest category (over 6) in the previous three maps. This diagram emphasizes the primacy of Berlin and the dominance of Germany in a network pattern more definitively expressed than in any of the previous analyses. There is not just a change in scale, there is a qualitatively different network: a German-speaking net of universities that almost alone define the great growth of European scientific practice. Outside this major network only one other subnet appears—the English triangle plus Edinburgh as a very minor part of modern science in nineteenth-century Europe. Overall Fig. 6 represents the origins of the university-based modern science that dominated worldwide scholarship in the twentieth century.

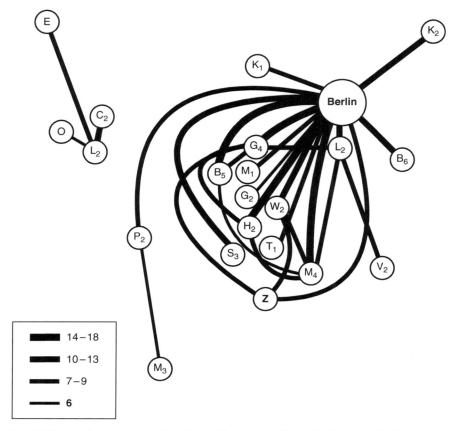

Fig. 6 Nineteenth-century networks of scientific practice. City codes: B_5 Bonn, B_6 Breslau, C_2 Cambridge, E Edinburgh, G_2 Giessen, G_4 Göttingen, H_2 Heidelberg, K_1 Kiel, K_2 Königsberg, L_2 Leipzig, L_3 London, M_1 Marburg, M_3 Montpellier, M_4 Munich, O Oxford, P_2 Paris, S_3 Strasbourg, T_1 Tübingen, V_2 Vienna, W_2 Würzburg, Z Zurich

A Theoretical Excursion into Town–Gown Conflict

University towns have a special place in the urban theory that is employed in this study. Active nodes in vibrant networks are expected to produce large successful cities (Jacobs, 1969; Taylor, Hoyler, & Verbruggen, 2008). But consider the following places that feature in the tables and maps above: Cambridge, Erlangen, Giessen, Göttingen, Heidelberg, Helmstedt, Jena, Marburg, Modena, Oxford, Pavia, St Andrews, Tübingen, Uppsala, Vienne, Wittenberg, and Würzburg. They are all small or medium-sized urban places[24] that have been on the demographic margins of the great urbanization revolutions that made Europe the most urbanized world region in the modern world-system by the end of the nineteenth century.

And yet these places have all been relatively important centers for the development of modern science. The conundrum is, therefore, that since science is so important to modernity and since the main demographic feature of modernity has been historically unprecedented high levels of urbanization, why have so many science centers *not* become major cities? Particular answers can be given for every case above, but given the quantity of cases, can there be a general explanation?

The first point to make is that there are also some places referred to in previous analyses that are today large cities—Berlin, London, and Paris are the obvious examples. But these exceptions prove the rule: They are places that have grown as multifunctional modern capital cities in which science practice has not been an overriding dimension. And that is the point. The 17 science centers listed above have all been dominated by the science workplaces that are universities. In Jacobs's (1969) theory of economic expansion through cities, the key process is diversification of the division of labor to create complex economic entities.[25] The converse of this is the "company town." Such settlements might be efficient from the company point of view, but their innate economic simplicity lessens opportunities for generating new work. Mill towns remain mill towns. Creating a new capital city is a similar one-function place production process that has created numerous politically powerful, but relatively small, places (e.g., Abuja, Brasilia, Canberra). It took Washington, D.C., nearly two centuries to become a major U.S. city because of its beginnings as a "company town"[26] (Abbott, 1999). In other words, large clusters of scientists (i.e., universities) condemn their towns to likely never becoming important cities. University towns are company towns; like other "companies," they thrive through monopolizing power in relatively small urban places at the expense of the economic expansion of those places.[27] Padua never became a Venice or Genoa, Leiden never became an Amsterdam, Uppsala never became a Stockholm, and Heidelberg and Würzburg never became a Hamburg or Munich. This monopolistic obstacle to economic growth is the essence of the town-versus-gown conflict.[28]

Conclusion

We have presented an extensive social science analysis of the "rise of modern science" and briefly revealed one local economic effect of the rise. Starting with a northern Italian Renaissance center of scientific practice based on Padua in the sixteenth century, the net expanded across the Alps to produce a Europe-wide polycentric network in which the scientific revolution blossomed. This integrated net dissipated to a large degree in the eighteenth-century Enlightenment, resulting in a disjointed network loosely centered on Paris. It was all change in the nineteenth century, with the German invention of the modern university harnessing scientific research and totally dominating advanced scientific practice. Taking the story forward, we can say that the Berlin-centered net expanded across the

Atlantic to produce a worldwide polycentric science network in the twentieth century. Extrapolation of the model suggests that this successful net's organization will dissipate in the current century. We might argue that this is happening as scientific research leaves the university for new corporate masters, constituting another qualitative change in the nature of scientific practice (Wallerstein, 2004a).

We began this chapter by setting out a world-systems frame that interweaves spatial core-periphery structures with knowledge structures, and we conclude with a comment on what this research has confirmed about our approach. The "rise of modern science" can be interpreted as an archetypal core-making process that has worked its way through the history of the modern world-system to become central to what "modern" is. But it has a particular geography, and this is not necessarily congruent with other core-making processes: There is no simple, spatially coherent bundle of core processes. The latter would imply a rather uniform core zone of social practices, which, of course, has never existed. The interesting divergence of the modern-science process from other core-making is to be found in the relative unimportance of Britain during the period of its hegemonic cycle (late-eighteenth century through the nineteenth century).[29] Britain was predominant in many things during its hegemony but not in "science." This relates to our interpretation of the town–gown conundrum: It was not in Oxford and Cambridge that new practical technologies were created; rather the great cities of northern Britain such as Manchester, Birmingham, and Glasgow were the vibrant cities underpinning British hegemony. Structures of knowledge are integral to the reproduction of the modern world-system in their own right. They do not simply mirror leading material and technological processes (Wallerstein, 2004b).

Notes

1. The term *structure* is used in this context to mean the slowly changing bases of the modern world-system from which processes emanate to constitute the historical system (Sayer, 1992; Wallerstein, 1974).

2. According to Wallerstein (2004a, p. 7), "For very many, the label 'scientific' and the label 'modern' became virtually synonymous, and for almost everyone the label was meritorious."

3. Ben-David used a core-periphery model in his classic studies of the rise of modern science (Schott, 1993, pp. 475–477). Our world-systems analysis differs from his work through setting these structures, and therefore scientific practice as a core-making process, within a historical systems framework (Wallerstein, 2004b).

4. Gascoigne has a very traditional approach to history that is "Whiggish" in nature: He uses twentieth-century categories to describe pre-twentieth-century practices in an evolutionary argument. However, we understand that concepts such as *scientist*, and indeed *science* and therefore *scientific career*, are quite problematic as descriptors for the time-scale of the modern world-system. For instance, Shapin and Thackray (1974) have written that "[t]o write the history of any period before ca 1870 primarily in terms of such unqualified modern categories is to endanger the enterprise at its inception [with] teleological assumptions" (p. 3). However, there are intellectual practices throughout the modern world-system that have come to be interpreted as contributing to the rise of what is now understood to be "modern science." These are the practices we are concerned with in this study and for which Gascoigne provides

relevant information in the form of systematically organized data. Thus, although we problematize the modern categories that describe the intellectual practices that are implicated in "the rise of modern science," we do argue there to have been a process that can be traced through the history of the modern world-system that we will categorize as *scientific practice*. These earlier practices are not precisely "modern science," for they occur in different social contexts; but they do represent a lineage of work leading up to modern science. It is this lineage which we shall call *scientific practice*, that we study in the present chapter. Obviously, just because "science" was not socially constructed as a concept until the late-nineteenth century, this does not mean that practices now viewed as "scientific" were not undertaken before such modern conceptualizations.

5. For more detail on Holland, see Davids (2001).

6. This is another way in which our study differs from the work of Ben-David. He uses national units of analysis without any systematic investigation of the urban places where science was practiced (Ben-David, 1971; Schott, 1993, pp. 458–462).

7. We appreciate that the author provides "only a brief and broad introduction" (p. 12), but it is in such limiting situations, where hard choices of inclusion and exclusion have to be made, that essential thinking is revealed.

8. Powell (2007) similarly takes a largely place-orientated perspective in his review of geographies of science within and beyond the discipline of geography but points out the potential for future work on movement and circulation.

9. Humanistic approaches to human geography were a reaction against studies of spatial models that reduced human beings to automatons (e.g., "economic man"). The critique involved replacing theories of space by meanings of place in which interpretation of people was much more complex and recognizably human. This has carried over into studies of scientific practice through prioritizing place over flows, but it is now changing. For example, Livingstone (2003) discusses "Circulation: Movements of Science" (pp. 135–178) as one of three key "geographical modalities" (p. 14) in his seminal exploration of geographies of science (the others being "Site: Venues of Science" [pp. 17–86] and "Region: Cultures of Science" [pp. 87–134]).

10. See note 5, above.

11. Wallerstein employs a critical realist methodology that encompasses two main approaches: intensive research and extensive research (Sayer, 1992). The former involves detailed study of the agents/actors who create the processes, whereas extensive research focuses on the broad patterns of what are usually quantitative data. Extensive research is often used as a prelude providing the statistical context for intensive research. This chapter is an exercise in extensive social science: Patterns of nodes and networks are described, but the detailed interpretation of the agents in these places and flows—a further step toward improving understanding of the "rise of science"—is not attempted here.

12. Castells uses the two spaces to argue that contemporary globalization is characterized by spaces of flows dominating spaces of places as the key social space. We follow Arrighi (1994), who shows that such an imbalance is not unique to the present; the concepts can be used as historical categories (see also Taylor, 2007).

13. The limiting cases of a purely fluid space of flows and a purely inert space of places do not exist in social relations (see Taylor, 2007).

14. See Allen's (1999) discussion on "city networks" versus "networks of cities".

15. Other relevant flows include academic travel. See, for example, Jöns (2008).

16. Of course, these sources are notoriously "Whiggish" in nature (see note 5), but nevertheless they do provide relevant information for deriving a sample of relevant individuals who have contributed to the "rise of modern science." In this way we employ a "collective biography" or basic prosopographic approach through aggregating "career" paths in knowledge production. This is the lineage of modern science described below. Like any empirical study, the results are only as good as the data. In this case we treat Gascoigne's encyclopedic work as a reasonable starting point, though recognizing that it could be improved. But that is for

another research effort; the credibility of the results presented below do strongly suggest that Gascoigne provides a reasonable initial basis for describing the lineage of modern science.

17. These centuries also broadly fit Wallerstein's (1974 1980, 1989) chronicling of the early modern world-system: Creation centered on the sixteenth century (reorientation of Mediterranean economy), consolidation centered on the seventeenth century (rise of North West Europe especially the Dutch), rivalry centered on the eighteenth century (mercantile struggles), and expansion centered on the nineteenth century (industrial revolution). Note also that Fig. 1 contrasts with Riddle's (1993) data on university foundings (Fig. 2, p. 55), which, as she points out (p. 55), show no relation to Wallerstein's world-system cycles. Clearly, scientific practice and the establishment of universities are distinctive and separate processes, the latter being particularly influenced by political structures (Riddle, 1993). For spatial patterns of university foundations from 1500 to 1800, see Frijhoff (1996, pp. 95–105).

18. Return to a previous workplace was also listed.

19. Because we focus on the rise of modern science in Europe, information on U.S. scientists (predominantly nineteenth century) was also excluded from the study.

20. A large body of literature on the subject has developed since Eulenburg's (1908) early study of academic recruitment in Germany. For an explicitly geographical perspective, see Meusburger (1990), for example.

21. One of the most obvious features of these maps is that they each have a wide distribution of scientists across Europe. And yet, during the four centuries they cover, developments in means of transport developed greatly, culminating in railways in the nineteenth century. But the nineteenth-century map (Fig. 2) has roughly the same spread, with just a little expansion to the east. We know from merchant activities that travel was Europe-wide by the sixteenth century, and scientists seem to have covered this same activity space throughout the periods included in this study.

22. A primate settlement pattern occurs when one center dominates—is much larger than—all the other places.

23. A polycentric settlement pattern is where there are several roughly equal centers. That is, no one place dominates. It is the opposite of a primate distribution.

24. All of these urban places have populations under 200,000 today.

25. Jacobs (1969, pp. 85–121).

26. Fifer (1981) actually referred to Washington, D.C., as "a company town."

27. Our study ends in 1900, but it can be noted that in several cases in the twentieth century "town" has been able to fight back successfully against "gown," turning, for example, Oxford into a major motor manufacturer and Cambridge into a high-tech center in the recent economic climate where universities are keen to show they are economic assets rather than obstacles: Spin-offs are demanded in return for high levels of state support. Historically, however, universities have been severe obstacles to economic growth. That is why so many centers of science practice in our geohistory are small places.

28. For an orthodox discussion of the town–gown conflict, see Brockliss (2000).

29. Hegemonic cycles constitute the historical frame of Wallerstein's (1984) modern world-system and are constituted by the rise, consolidation, and fall of a power possessing dominant economic, cultural, and political power (i.e., state hegemony). Wallerstein identifies three such cycles, Britain's hegemony occurring after the Dutch and before American hegemony.

References

Abbott, C. (1999). *Political terrain: Washington, D.C., from tidewater town to global metropolis.* Chapel Hill: University of North Carolina Press.

Allen, J. (1999). Cities of power and influence: Settled formations. In J. Allen, D. Massey, & M. Pryke (Eds.), *Unsettling cities* (pp. 181–228). London: Routledge.

Arrighi, G. (1994). *The long twentieth century.* London: Verso.

Ben-David, J. (1971). *The scientist's role in society: A comparative study.* Englewood Cliffs, NJ: Prentice-Hall.

Ben-David, J., & Zloczower, A. (1962). Universities and academic systems in modern societies. *Archives européennes de sociologie, 3,* 45–84.

Brockliss, L. (2000). Gown and town: The university and the city in Europe, 1200–2000. *Minerva, 38,* 147–170.

Castells, M. (1996). *The information age: Vol. 1. The rise of the network society.* Oxford, England: Blackwell.

Davids, K. (2001). Amsterdam as a centre of learning in the Dutch Golden Age, c. 1580–1700. In P. O'Brien, D. Keene, M. 't Hart, & H. van der Wee (Eds.), *Urban achievement in early modern Europe* (pp. 305–325). Cambridge, England: Cambridge University Press.

Eulenburg, F. (1908). *Der "Akademische Nachwuchs". Eine Untersuchung über die Lage und die Aufgaben der Extraordinarien und Privatdozenten* [The next generation of scholars: A study of the situation and tasks of the associate professors and unsalaried university lecturers]. Leipzig: Teubner.

Fifer, J. V. (1981). Washington, DC: The political geography of a federal capital. *Journal of American Studies, 15,* 5–26.

Frijhoff, W. (1996). Patterns. In W. Rüegg (Series Eds.) & H. de Ridder-Symoens (Vol. Ed.), *A history of the university in Europe: Vol. II. Universities in early modern Europe (1500–1800)* (pp. 43–110). Cambridge, England: Cambridge University Press.

Gascoigne, R. (1984). *A historical catalogue of scientists and scientific books from the earliest times to the close of the nineteenth century.* New York: Garland.

Gascoigne, R. (1987). *A chronology of the history of science, 1450–1900.* New York: Garland.

Gascoigne, R. (1992). The historical demography of the scientific community, 1450–1900. *Social Studies of Science, 22,* 545–573.

Granovetter, M. S. (1973). The strength of weak ties. *American Journal of Sociology, 78,* 1360–1380.

Habermas, J. (1989). *The structural transformation of the public sphere.* Cambridge, MA: MIT Press.

Harris, S. J. (2006). Networks of travel, correspondence, and exchange. In D. C. Lindberg & R. L. Numbers (Series Eds.) & K. Park & L. Daston (Vol. Eds.), *The Cambridge history of science: Vol. 3. Early modern science* (pp. 341–362). Cambridge, England: Cambridge University Press.

Jacobs, J. (1969). *The economy of cities.* New York: Vintage.

Jöns, H. (2008). Academic travel from Cambridge University and the formation of centres of knowledge, 1885–1954. *Journal of Historical Geography, 34,* 338–362.

Livingstone, D. N. (2003). *Putting science in its place: Geographies of scientific knowledge.* Chicago: University of Chicago Press.

Livingstone, D. N. (2005). Text, talk and testimony: Geographical reflections on scientific habits— An afterword. *British Journal for the History of Science, 38,* 93–100.

Lux, D. S., & Cook, H. J. (1998). Closed circles or open networks?: Communicating at a distance during the scientific revolution. *History of Science, 36,* 179–211.

Meusburger, P. (1990). Die regionale und soziale Rekrutierung der Heidelberger Professoren zwischen 1850 und 1932 [The regional and social origin of Heidelberg's professors between 1850 and 1932]. In D. Barsch, W. Fricke, & P. Meusburger (Series Eds.) & P. Meusburger & J. Schmude (Vol. Eds.), *Heidelberger Geographische Arbeiten: Vol. 88. Bildungsgeographische Studien über Baden-Württemberg* (pp. 187–239). Heidelberg: Selbstverlag des Geographischen Institut der Universität Heidelberg.

Naylor, S. (2005). Introduction: Historical geographies of science—Places, contexts, cartographies. *British Journal for the History of Science, 38,* 1–12.

Powell, R. C. (2007). Geographies of science: Histories, localities, practices, futures. *Progress in Human Geography, 31,* 309–329.

Riddle, P. (1993). Political authority and university formation in Europe, 1200–1800. *Sociological Perspectives, 36,* 45–62.

Sayer, A. (1992). *Method in social science: A realist approach* (2nd ed.). London: Routledge.

Schott, T. (1993). The movement of science and of scientific knowledge: Joseph Ben-David's contribution to its understanding. *Minerva, 31*, 455–477.

Shapin, S. (2006). The man of science. In D. C. Lindberg & R. L. Numbers (Series Eds.) & K. Park & L. Daston (Vol. Eds.), *The Cambridge history of science: Vol. 3: Early modern science* (pp. 179–191). Cambridge, England: Cambridge University Press.

Shapin, S., & Thackray, A. (1974). Prosopography as a research tool in history of science: The British scientific community 1700–1900. *History of Science, 12*, 1–28.

Taylor, P. J. (2007). Space and sustainability: An exploratory essay on the production of social spaces through city-work. *The Geographical Journal, 173*, 197–206.

Taylor, P. J., Hoyler, M., & Verbruggen, R. (2008). *External urban relational process: Introducing central flow theory to complement central place theory* (Globalization and World Cities Research Network [GaWC] Research Bulletin No. 261). Loughborough, England: Loughborough University.

Thrift, N. (1999). Cities and economic change: Global governance? In J. Allen, D. Massey, & M. Pryke (Eds.), *Unsettling cities* (pp. 271–320). London: Routledge.

Wallerstein, I. (1974). *The modern world-system: Capitalist agriculture and the origins of the European world-economy in the sixteenth century*. New York: Academic Press.

Wallerstein, I. (1980). *The modern world-system II: Mercantilism and the consolidation of the European world-economy, 1600–1750*. New York: Academic Press.

Wallerstein, I. (1984). *The politics of the world-economy*. New York: Academic Press.

Wallerstein, I. (1989). *The modern world-system III: The second era of the great expansion of the capitalist world-economy, 1730–1840s*. New York: Academic Press.

Wallerstein, I. (1999). *The end of the world as we know it: Social science for the twenty-first century*. Minneapolis: University of Minnesota Press.

Wallerstein, I. (2004a). *The uncertainties of knowledge*. Philadelphia: Temple University Press.

Wallerstein, I. (2004b). *World-systems analysis: An introduction*. Durham, NC: Duke University Press.

From Mediocrity and Existential Crisis to Scientific Excellence: Heidelberg University Between 1803 and 1932

Peter Meusburger and Thomas Schuch

Universities are sometimes idealistically seen as an institutional embodiment of academic ideas, as prominent places of research where universally valid scientific truths are generated and taught, or as havens of intellectual freedom. From the outset, however, universities have been shaped by political, economic, religious, and ideological interests. They have depended on external financial resources, suffered from internal and external conflict, interacted with their cultural environment, and derived part of their intellectual vigor from the extent and character of their spatial relations. One can use many diverse approaches to investigate the intellectual ups and downs of universities, drawing on several indicators of research output and scientific reputation. Most studies on the historical development of universities, however, focus on the individual biographies of eminent scholars and the impact they have on departments and disciplines.

This chapter pursues a different strategy. First, we intend to complement biographical (and often qualitative) studies on individual scholars by providing a kind of structural framework to which individual biographies can be related. Second, we try to combine an idiographic description of the history of Heidelberg University with a nomothetic approach for analyzing general trends and structures of the German university system. Third, we want to understand how complex relations between various levels of aggregation interact. We seek to avoid the traps of both the individualistic and holistic approaches. Like Best (1981) and Weick (1995), we are interested in the dynamics of collective career patterns and staff structures. With this approach we make inferences from the typical or general aspects of a given historical and spatial context without neglecting its atypical, aberrant, or individual elements.

P. Meusburger (✉)
Geographisches Institut, Universität Heidelberg, Heidelberg 69120, Germany
e-mail: peter.meusburger@geog.uni-heidelberg.de

T. Schuch
Geographisches Institut, Universität Heidelberg, Heidelberg 69120, Germany
e-mail: thomas.schuch@geog.uni-heidelberg.de

P. Meusburger et al. (eds.), *Geographies of Science*, Knowledge and Space 3,
DOI 10.1007/978-90-481-8611-2_4, © Springer Science+Business Media B.V. 2010

Universities differ in the scientific reputation of their professors, the quality of their students, the expensiveness of their research infrastructure, the density and range of their international networks, the degree of their academic visibility, and the array of their graduates' career prospects. The resulting rank order of universities is represented in a hierarchy of places. This spatial hierarchy has a factual side (e.g., uneven spatial distribution of financial resources), a socially constructed side (e.g., uneven distribution of scientific reputation), and an affective side (scholars' possible emotional affinity with some places and aversion to other places). In this chapter we take Heidelberg University as the focus for an analysis of factors contributing to the intellectual vibrancy of universities. Our aim is to show that universities are embedded in regulatory contexts, and we argue that the functioning, complexity, and intellectual ebb and flow of a university cannot be understood exclusively through biographies of distinguished individual scholars. We appreciate factors like the importance of favorable financial circumstances, new organizational structures, broad academic autonomy, advantageous career paths, and new policies on faculty appointments. However, we also emphasize that none of these factors alone could have catapulted Heidelberg University out of an existential crisis into the top group of European universities within only four or five decades. These factors do not act in isolation; they interact, reinforcing or modifying each other as a kind of macrophenomenon called "setting," "milieu," or "spatial context." We intend to show that the generation and circulation of scientific knowledge is affected by such context, environment, and spatial relations.

We define spatial context as interrelated social, cultural, political, and material conditions in which institutions exist and scholars act. A context is a phenomenon of emergence. It is more than the aggregation of single causes; it is a kind of social space that is process-oriented, dynamic, and subject to reinterpretation and modification. It facilitates or constrains social relations and actions. A context is more than a stage on which actions take place, however it is not an active venue that shapes the practices of knowledge acquisition and circulation. Instead, it is a volatile potentiality utilized by some scholars and disregarded by others. A context can only be defined and evaluated in relation to specific agents. Certain contexts and sites "dictate what we can say and do in particular circumstances and—just as important—what we can't. Every social space has a range of possible, permissible, and intelligible utterances and actions: things that can be said, done, and understood" (Livingstone, 2003, p. 7).

A context entails more than just social relations and networks; it also includes the materiality and accessibility of objects. For instance, expensive research facilities such as laboratories are not ubiquitously available. Certain categories of knowledge can be generated only at certain places (sometimes under restricted access) that also have an impact on the circulation of ideas and scholars. Some factors of the context are spatially rooted and take considerable time to change (e.g., elaborate research infrastructure). Some of them are clearly linked to certain administrative units with defined borders but can swiftly be modified (e.g., state subsidies to universities). Other factors exist mainly through social relations as patterns of interactions or arrangements of power and can be altered at short notice by the actors involved. Metaphorically, a place arguably catenates the influences operating on multiple scales and funnels them into a locally effective context. Whether a milieu is detrimental or conducive to an activity depends on the intentions, motivations, resources, and competence of an actor or social system interacting with it (see Meusburger, 2008, 2009).

A creative milieu sets off a self-reinforcing upward spiral of academic respectability. Outstanding scholars are more likely to accept an offer from a university if other such scholars are already employed there. As a rule, a faculty of renowned scholars applies higher standards in its appointments, has closer personal relations to leading scholars of its disciplines, is far more attractive to potential applicants, and thus can choose from a larger pool of candidates than is the case at an academic institution of less repute.

The Intellectual Decline of Heidelberg University

Since 1386, the year in which Heidelberg University was founded, the institution has experienced many rises and falls in its scientific prestige, intellectual influence, and international appeal. Its first golden age came in the second half of the sixteenth century and the first two decades of the seventeenth century, when it was a center of humanism and a stronghold of Calvinism. At that time it flowered intellectually and exerted major influence throughout Europe. After 1622 Heidelberg University suffered from disruptions in teaching due to the 30 Year's War. The university lost its illustrious Palatine Library (*Bibliotheca Palatina*), which was transported as booty to the Vatican in 1623. These conflicts of the Counterreformation and the Palatinate War of Succession (1688–1697) eventually resulted in the total destruction of the town and all its university buildings (1689 and 1693). The final descent into mediocrity took place in the eighteenth century, when the university fell victim to the Counterreformation's long aftermath. The prevailing authorities at that time required the university faculty to be Catholic and restricted the institution's autonomy in other ways, with the government monitoring instruction to an ever greater degree.

The intellectual deterioration of Heidelberg University was accelerated by financial crises. In the second half of the eighteenth century, the university's budget was no longer sufficient to interest scholars of repute, and from 1780 onwards the institution's expenditures regularly exceeded its income. According to Wolgast (1986a, p. 83), the financial situation became hopeless when the Napoleonic wars engulfed the areas west of the Rhine. When they passed to France under the peace agreements of Campo Formio (1797) and Lunéville (1801), the university ceded a good deal of its properties and income. The ensuing low salaries meant that many professors depended on secondary sources of income, a need that reduced their rates of attendance and the quality of their teaching. Like many European universities in the eighteenth century, Heidelberg University had a poor scientific reputation. It was marked by decay, inefficiency, and stagnant intellectual life. Rulers of those times saw study abroad as a drain on the economy, so they required their subjects to study at home. In short, Heidelberg University had lost any international or national role it once had and served mainly local interests. Most of its professors were appointed in a way that had little to do with notable scholarly reputation. The university suffered from the not uncommon eighteenth-century practice of bequeathing chairs to sons, sons-in-law, or grandsons and of giving preference to local candidates (for details see Wolgast, 1986a, pp. 77–78, 1986b, 2008). The academic nadir reached by the late eighteenth century was well noted by contemporaries. Friedrich Gedike, visiting the German universities on behalf of Prussian King Frederick William II, for example, wrote of Heidelberg in 1789: "Everything that I saw and heard convinced me that this university is insignificant" (cited in Wolgast, 1986a, p. 84, our translation). In 1798 the rector of the university reported that "the University of Heidelberg displays the infirmities of advanced old age: dullness and inactivity" (cited in Wolgast, 1986a, p. 5, our translation). In 1798 the financial plight of the university prompted a government commission to conclude that the institution had to be deemed "a terminally ill patient ... whom it would be best to allow to die in peace" (cited in Mussgnug, 2003, p. 131, our translation).

Another threat to Heidelberg University's survival was its small size. Between 1620 and 1700 its average annual enrollment of 109 students placed it 24th among the 26 German universities (Eulenburg, 1904, p. 85). Between 1700 and 1790 it held 22nd place among 31 universities (Eulenburg, 1904, p. 153). The dissolution of ecclesiastical states and scores of small political units in Germany (1803) sounded the death knell for many of these small universities or at least interrupted their business for a time. Between 1786 and 1818, 22 of a total of 42 universities in the territory of the Holy Roman Empire of the German Nation were closed (Weisert, 1983, p. 71). When the Palatinate Electorate was dissolved in 1802 and Heidelberg acceded to the state of Baden in the course of Napoleon's territorial reorganization of German lands, Heidelberg University was faced with bankruptcy. It was 6 months in arrears on its payments to professors and owed them an entire year's payment in kind, mostly in the form of wood and wine (see Wolgast, 1986a, p. 87). If Heidelberg University had not been set on a new financial footing by Margrave (later Grand Duke) Carl Frederick of Baden, and if it had not been for the fundamental reforms of 1803, its existence would most likely have ended.

Factors in the Rise of Heidelberg University in the Nineteenth and Early Twentieth Centuries

Affiliation to a Particular Administrative Unit

In a realm as fragmented as Germany was in the nineteenth century, the politicoadministrative affiliation of a university was crucial to its development and academic prospects. In each state or territory, the bodies responsible for its universities endowed them with various financial resources, organizational structures, and degrees of academic autonomy and political freedom. Fortunately for Heidelberg University, the town of Heidelberg, together with other municipalities from the dissolved Palatine electorate, acceded in 1802 to the Margravate of Baden (which became the Grand Duchy of Baden in 1806) in the course of the territorial reorganization initiated by Napoleon. At that time Baden had no university and for sundry reasons had an interest in taking over the University of Heidelberg and encouraging its development. Baden's acquisition of its own university (Freiburg did not become part of Baden until 1805) was not only a question of enhancing margravate's political stature. It was also about precluding excessive foreign influence on the education of the senior civil servants and free professions serving the state and of guaranteeing the clergy's theological orthodoxy. In the context of enlightened absolutism, the main aims of university reforms were to modernize the teaching at the universities so that new and useful ideas were not neglected; to increase the military, fiscal, and economic efficiency of the territorial state; and to extend the state's productive capacities. Margrave (later Grand Duke) Carl Frederick gave Heidelberg University a new organizational structure between 1803 and 1805, and the Grand Duchy took over the university's funding, providing it with fiscal security and calculability (for details see Wolgast, 1986b, 2000, 2008).

Financial Resources

The budget of Heidelberg University grew considerably in the course of the nineteenth century. From 1850 to 1914, the Grand Duchy of Baden invested heavily in its universities and other scientific institutions and appointed top academics to its universities. In that period Baden committed much more of its state budget to universities than did Prussia or any other German state (Table 1).

Table 1 Expenditure for science in Baden and Prussia

Years	Expenditure for science in percentages of the state budget		Per-capita expenditure for science (converted into German marks)	
	Baden	Prussia	Baden	Prussia
1850–1859	1.9	0.8	0.42	0.15
1860–1869	2.3	0.7	0.40	0.17
1870–1879	4.0	1.6	0.83	0.41
1880–1889	4.4	1.8	1.01	0.56
1890–1900	4.3	1.0	1.53	0.60
1900–1909	4.4	1.2	1.82	0.97
1910–1914	4.7	1.1	2.20	1.22

Note: From Pfetsch (1974). Reprinted and adapted with permission of Duncker & Humblot.

From 1870 to 1914, Baden invested between 4.0 and 4.7% of its state budget[1] in its universities and other scientific institutions, spending more than twice that amount on them per capita as Prussia did on its universities. In the second half of the nineteenth century, Baden consciously used its universities to promote industrial development and turned to science to reinforce the state's economic strength and status. These policies stood Heidelberg University in good stead, for the second industrial revolution, which occurred in Germany in the last three decades of the nineteenth century, was driven by research-intensive branches such as chemistry and electrical engineering. It was a time when science itself and research-driven universities were rapidly developing. The natural sciences and medicine, which were costly, especially benefited from the generous salaries of professors and from the expensive laboratories Baden offered to leading academics. In order to attract the famous chemist Robert Bunsen (1811–1899) from Breslau to Heidelberg, the Baden government funded a state-of-the-art chemistry department, the most modern in Germany. Bunsen earned up to 72% more than his predecessor (including a housing allowance) and had the second highest salary of all professors in Heidelberg (Wolgast, 1986a, p. 103). Bunsen was a prime reason why other exceptional scientists, including Hermann Helmholtz (1821–1894), Robert Kirchhoff (1824–1887), and Leo Koenigsberger (1837–1921) joined the staff of Heidelberg University and created a research center of world-wide reputation.

University Policy—Autonomy of Universities

Baden's ministerial bureaucracy allowed Heidelberg University broad autonomy. In the early nineteenth century, however, Heidelberg University could have taken quite a different direction. At that time, two opposing concepts of the nature and goals of a university were competing in Baden's administration, associated with the names Sigismund von Reitzenstein (1766–1847) and Johann Niklas F. Brauer (1754–1813; for a detailed account on both men, see Wolgast, 1986a, 1987, 2008). Reitzenstein, influenced by his studies in Göttingen, was an adherent of the modern, neohumanist understanding of the university and argued against excessive state regulation, which in his opinion would kill the free, living spirit of research. His role model was apparently Gerlach A. von Münchhausen (1688–1770), who had made the University of Göttingen (founded in 1737) the most innovative and prestigious German university of the time. Its modernity and sterling reputation were largely due to the unusual freedom granted to the professors and to the fusing of teaching with research functions. "Göttingen's freedom to think, write, and publish was unsurpassed in Germany" (McClelland, 1980, p. 39). Professors were free from close religious supervision, and their chief responsibilities were to advance knowledge and to carry out original research. The system of knowledge taught in Göttingen was no longer based on theology, church history, canon law, or Roman law but rather on what were termed the "state sciences", including history, geography, statistics, economics, and modern languages. Göttingen University was financed by the state and did not depend on endowments.

Reitzenstein was opposed by Brauer, who represented the ideals of the late absolutist state and enlightened pragmatism. Brauer saw the university primarily as an educational institute for the training of civil servants, a responsibility that should be strictly controlled. He wished to limit professors' freedom to teach as they saw fit and expected them to promulgate only uncontested and generally recognized knowledge, basing their courses on approved textbooks. In 1804 he ruled that all persons seeking a career in the civil service or the church in Baden had to study in Baden. After all, he argued, the precepts according to which its civil servants were educated could not be a matter of indifference to the state (for details see Wolgast, 1987, p. 41; 2008, p. 13). Reitzenstein ultimately prevailed with his neohumanist idea of the university. In his view, the university was not merely a school for transmitting uncontested knowledge or training loyal civil servants but a place where scientific research could be freely pursued. He and his successors Wilhelm Nokk (1832–1903) and Franz Böhm (1861–1915) respected the autonomy of the universities and valued freedom in research and teaching. Reitzenstein cooperated well with Heidelberg University, never abusing his appointive power for nonscholarly purposes. All these facts were critical to the academic advancement of Heidelberg University. The staffing and appointments policies of the faculties and the concomitant intellectual renewal and openness to modern scientific trends played an equally central role.

This example shows the importance of serendipity and of interdependencies between the micro- and macro-levels. If Brauer had succeeded, or if Reitzenstein had not tread Göttingen's path, Heidelberg University would have developed quite differently. By the same token, the route Göttingen took could have been irrelevant if some of the influences discussed in the rest of this chapter had not contributed to Heidelberg's milieu.

Freedom of Thought and Political Climate

Another key factor of Heidelberg University's comparative appeal was the political climate in Baden as a whole and in Heidelberg in particular. In the nineteenth century, Heidelberg University developed into a cosmopolitan and liberal university in most ways. In this context the term liberalism referred to a belief in freedom of thought and tolerance as well as to political constitutionalism. For example, Heidelberg's professors and students were actively involved in various movements for more freedom, democracy, institutions of civil society, freedom of the press, and national community. They attended the Wartburg Festival in 1817, the giant liberal demonstration at Hambach in the Palatinate in 1832, the abortive storming of the Frankfurt Guard in 1833, and the revolution of 1848. In the nineteenth century, many German students and professors moved to the liberal south, where they were subject to less surveillance and persecution and where publications were not as strictly censored as in other parts of Germany (Eulenburg, 1904, p. 186). Because Heidelberg University was seen as a bastion of German liberalism, Prussian students were forbidden to study in Heidelberg between 1833 and 1838 (Wolgast, 1986a, p. 99). This liberal climate was also appreciated by foreign students, some of them being political radicals. In the second half of nineteenth century, Heidelberg became a center for Russian revolutionaries.[2] In 1866 a czarist fact-finding committee announced that the purpose of the Russian library (*Lesehalle*) in Heidelberg was mainly to diffuse revolutionary propaganda (Walter, 2003, p. 211). From 1868 to 1881 between 20 and 29% of Heidelberg's students were foreigners, and in the 1890s that group occasionally accounted for as much as 35% of the students in the natural sciences. At that time the share of foreign students was far smaller at the three competing universities—up to 16% in Berlin, 15% in Leipzig, and 12% in Munich (Titze, 1995, pp. 46, 312–313).

Another indication of Heidelberg University's liberalism is the fact that the institution was one of the first German universities to appoint Jews as professors and librarians and to enroll women (as was the case in 1869 with the Russian Sofja Kovalevskaja, who later became professor of mathematics in Sweden).[3] According to McClelland (1980, p. 316), Baden was a haven for professors who were too controversial for other German states. The eminent sociologist Max Weber, for instance, described Baden's university policy as follows: "I freely admit that when I moved from an educational system under Prussian administration to Baden [in 1897], it felt like fresh air" (Weber, 1926/1984, p. 434, our translation). The regional culture and provincial politics of Baden conditioned the practices and products of scientific endeavors and let certain forms of scientific activity emerge at Heidelberg University as described by Livingstone (2003, p. 15).

Transdisciplinary Networks and Intellectual Milieu

In a small city like Heidelberg, the daily life of academics concentrates on a few streets, so they have little chance to evade each other. However, daily eye contact on the streets neither gives rise to friendships nor abolishes ignorance of and prejudice against methods and theoretical approaches of other disciplines. Spatial proximity of renowned scholars in a small town does not automatically stimulate communication or creative processes. The local potential for transdisciplinary discourse has to be activated and cultivated either through personal networks or through scientific circles and *jours fixes*. The friendly relations between Bunsen, Kirchhoff, Helmholtz, and Koenigsberger are an example of a personal network (for details see Koenigsberger, 1919).

As for scientific circles, Heidelberg was fortunate in the first third of the twentieth century to have at least six whose influence reached well beyond Heidelberg: Eranos, Weber, George, Thode, Gruhle, and Janus (for details see Allert, 1995; Breuer, 1995; Essen, 1995; Jellinek, 1970; Kolk, 1995; Sauerland, 1995; Treiber, 1995; Weber, 1984). Some scholars took part in more than one of them. The Eranos circle (1904–1912) focused on religious studies and comprised influential professors of archaeology, classical philology, economics, criminal law, pathology, philosophy, religion, and theology. The most famous was the *jour fixe* of Max and Marianne Weber (founded in 1911) and Alfred and Marianne Weber (1924), where (from today's perspective) celebrated scholars of anatomy, cultural history, economics, history, law, philosophy, psychiatry, sociology, and theology met first biweekly and eventually on an irregular basis at 17 Ziegelhäuser Landstrasse. The Eranos and Weber circles drew leading minds of the time (among them Hermann Braus, Martin Dibelius, Eberhard Gothein, Friedrich Gundolf, Karl Jaspers, Walter Jellinek, Emil Lask, Karl Löwenstein, Gustav Radbruch, Otto Regenbogen, Arthur Salz, Ernst Troeltsch, and Alfred Weber). Scholars from other universities, too, occasionally came, such as the economist Werner Sombart (Berlin), the sociologist Georg Simmel (Berlin), and the German-Italian sociologist Robert Michels.

These groups were also open to recommended students, such as Ernst Bloch, Paul Honigsheim, György Lukács, Edgar Salin, Ernst Toller, and Theodor Heuss (the first president of the Federal Republic of Germany). The myth of Heidelberg also influenced many doctoral and postdoctoral students not present in these circles (among them Robert E. Park, Karl Mannheim, Talcot Parsons, Hannah Arendt, and Norbert Elias). Because of this intellectual milieu, Heidelberg was known as the "world village" (Jellinek, 1970, p. 85), "Noah's ark" (Karádi, 1995, p. 378), and the "secret capital of intellectual Germany" (Sauerland, 1995, p. 12; see also Karádi, 1995, p. 378). Noah's ark alluded to the multiplicity of personalities and inspiring academic subcultures. The characterization as a world village referred to the vast area from which Heidelberg's professors had been recruited and to the sizeable proportion of the student body accounted for by foreigners hailing from Russia, Poland, Hungary, the United States, Great Britain, and numerous other countries before World War I (1914–1918). These catchwords capture the typical "Heidelbergian spirit" of the knowledge enterprises that parts of the university were engaged in during the first three decades of twentieth century.

Not only were the meeting places of these circles topoi of intellectual exchange or fora of scholarly challenges; they constituted new interactions between disciplines. They facilitated and conditioned "discursive space" (Livingstone, 2003, p. 7) and expressed a kind of scholarly exclusiveness. The Weber and Eranos circles exerted considerable power when it came to important faculty decisions in the 1910s and 1920s and the renewal of Heidelberg University after 1945. In Max Weber's opinion science policy was a public matter (Sauerland, 1995, pp. 15–16). Because of his intellectual authority and networks, he figured in many procedures bearing on the *Habilitation*, the German university's postdoctoral degree, or *venia legendi* (the license to give accredited lectures in a certain discipline at the university)[4]— and on recruitment procedures at other faculties and universities. Ludwig Curtius referred to Weber as "a dictator of an intellectual empire" (Essen, 1995, p. 462). Almost all the members of the Eranos circle were affiliated with the newly founded Heidelberg Academy of Sciences in 1909. Six of the members of the "Board of the 13,"[5] which was responsible for the denazification and renewal of Heidelberg University between 1945 and 1948, had been members of the Eranos or Weber circles.

Social Origin of Heidelberg's Professors

Social Opening of German Universities in the Nineteenth Century

The staffing of universities is influenced by a number of societal and economic trends, among them the social opening or closing of universities. Between 1803 and 1932 the social background of Heidelberg's professors underwent marked shifts that mirrored various societal, economic, and scientific trends. Because these shifts diverged from those of the university's students, a few comments about the social origin of the students seem necessary before we delve into the social origin of the professors. Compared to universities in other European nations, those in Germany in the eighteenth and early nineteenth centuries had a relatively large share of students from the lower and middle classes (for details see Conrad, 1884; Jarausch, 1984). One of the main reasons for making universities accessible to these strata of society was recruitment to the clergy. Intelligent sons of peasants were encouraged by the local priests to enter a grammar school and, later, a university. Both Protestant and Catholic universities had a system of providing scholarships, dormitories, and subsidized meals (see Anderson, 2004, p. 16). Some of Germany's most noted nineteenth-century intellectuals came from poor families. Students from such backgrounds gravitated first to the church or to teaching, both of which could provide a living as soon as final examinations had been passed—unlike public service, the judiciary, or universities, which usually entailed a long period of minimal income requiring continued family support (Anderson, 2004, p. 129).

Until 1870, German universities were dominated by students of the *Bildungsbürgertum*, a distinctive learned and cultivated stratum of the upper-middle class, which was "a kind of substitute nobility, an aristocracy of mind" (Ringer, 1979, p. 74). With the rapid swell in student numbers and the increase of transit quota to universities in the last decades of the nineteenth century, however, the proportion of nobles among students and the proportion of students whose fathers held a university degree (e.g., Table 2) declined notably.

Table 2 Social background of students in Württemberg

Period	Proportion of students whose fathers held a university degree (%)
1871–1876	43.41
1876–1881	35.97
1881–1886	29.62
1886–1891	35.25
1891–1896	34.35
1896–1901	34.21
1901–1906	31.67
1906–1911	28.77

Note: From Rienhardt (1918, p. 19). Reprint with permission of Mohr.

Family Background of Professors

In the eighteenth and early nineteenth centuries the standing of university professors was very low. Most universities were suffering from public opprobrium, mockery, a crisis of confidence, and a lack of funding. Shortly before the French Revolution, literati of the German Enlightenment mounted sweeping attacks on the universities, and many observers called for the outright abolition of these institutions (McClelland, 1980, p. 27; Turner, 1987, p. 223). In the early eighteenth century most university professors did not produce new knowledge, and the primary function of the university was not to produce learned men of science or professional experts but rather to bring forth men who were sociable, cultivated, and conscious of their social and moral worth. Not until the nineteenth century did the general trends toward meritocracy, professionalization, and bureaucratization turn universities into places of social selection that legitimated and, through examinations and academic degrees, controlled the entry into specific professions and the access to social privileges (see Anderson, 2004, pp. 13–14; McClelland, 1980, p. 31).

From the 1840s on, the image of professors in Germany improved rapidly. By the late nineteenth century "the German university system was the most admired in the world" (McClelland, 1980, p. 2), and university professors were viewed as the elite among the educated classes. According to McClelland, the impact of German universities on society was surely deeper and the universities' monopoly over access to the professions was much stronger than that of their American or British counterparts. In no other country was academic training so fundamental a prerequisite for high office as in the German states. The men who shaped Germany's cultural and scientific life were closer to universities than in most other parts of Europe and in the United States. German universities served as the breeding ground for the *Bildungsbürgertum*, the recruiting pool for both cultural and administrative elites (pp. 2–3, 7). The rising esteem of university professors made this career in the second half of the nineteenth century attractive also to the aristocracy and the commercial and industrial upper classes, enriching the fund of talent.

A methodological complication confronting most historical studies on the social origin and the vertical social intergeneration mobility of university professors is that they must rely on existing job titles mentioned in biographical sources. Most authors of such studies are not in the position to develop their own classification of occupations, as done in the empirical surveys by Eulenburg (1908) and Busch (1959). Likewise, we turn to the occupational titles mentioned in the short biographies of the *Gelehrtenlexikon* of Heidelberg University covering the years from 1803 to 1932 (Drüll, 1986), which states fathers' occupations for 689 of the 722 professors (95.5%). It is therefore necessary to accept titles such as trader or merchant, which defy attribution to a specific class and include a broad range of income levels. We point out, too, that our source does not state explicitly whether fathers were aristocrats, most of the nobles presumably having been registered as landowners, military officers, or senior government officials. Lastly, both the number and nature of occupations changed over the 130 years covered by this study. Despite these problems, though, an analysis of the father's occupation offers interesting insights into the recruitment of professors and the competition in academic careers.

In this part of the chapter, we focus on the 689 professors for whom we know the fathers' occupations and about whom we know that they were employed at the University of Heidelberg between 1803 and 1932. Based on this statistical population, our calculations show that the majority of the professors there in those years were from the upper or upper-middle class. Only about 13% of them came from the ranks of junior civil servants, artisans, laborer, peasants, and innkeepers (see Table 3).

If the proportion of the student body represented by students from the lower-middle classes rose discernably at Heidelberg University from the early nineteenth century to World War I, that of its professors with those origins just as plainly did the opposite. One reason why graduates from the lower social classes often chose not to embark on a career as an unsalaried university lecturer (*Privatdozent*) was the paltry remuneration on the first career steps as a lecturer or associate professor (*Extraordinarius*). At some universities graduates were appointed as lecturers only if they could demonstrate sufficient wealth to tide them over until their first appointment as professor. A university lecturer had no fixed salary, and the income from lecture fees might come to only 5% of that earned by a full professor, or *Ordinarius* (for details see Turner, 1987, p. 233). Only those scholars who did not depend on a regular salary could afford to occupy such a position. The Heidelberg philosopher Hermann Glockner, who earned his *Habilitation* at that university in 1924, was told by the philosopher Heinrich Rickert, "You know that the faculty can only allow young men to complete their *Habilitation* if they have made a written commitment not to require even the smallest amount of financial assistance or support because they are able to support themselves from their own private means" (Glockner, 1969, p. 180, our translation).

Another reason that the share of the lower social strata receded among professors at German universities in the second half of the nineteenth century was the intensified competition for such posts, for by that time the social respectability of an academic career had appreciated considerably. Universities had come to be perceived as an "important symbol of German cultural achievements" (Turner, 1987, p. 221), with professors being regarded as the elite among scholars. This new appeal had enlarged the number of candidates from the upper classes (Table 3). The proportion of sons of senior civil servants among Heidelberg's professors surged from the period before 1830–1890, yielding only slightly in the 20 years thereafter. The share represented by the sons of manufacturers, industrialists, bankers, and landowners grew nearly 16-fold over the same span, and that of the sons of merchants, traders, and entrepreneurs more than doubled in the 130 years under study. The direct inheritance of status—reflected by the proportion of Heidelberg's professors accounted for by the sons of university professors, scientists, and other scholars—subsided only slightly during that period, whereas the sons of teachers nearly doubled their representation by 1910 and almost doubled it again by 1932. The group that contracted the most among the professors at Heidelberg University in the nineteenth century consisted of the clergy's sons, whose share

Table 3 Family background of professors employed at Heidelberg University, 1803–1932

Father's occupation	First appointment at Heidelberg University						Total	
	Before 1830	1831–1850	1851–1870	1871–1890	1891–1910	1911–1932	%	N
University professor, scientist, or other kind of scholar	15.4	10.5	11.7	6.9	14.3	10.5	11.6	80
Senior official, higher civil servant, judge, attorney, lawyer, doctor, apothecary, politician, public prosecutor, commissioned officer	19.8	28.9	27.3	33.3	27.3	26.8	27.1	187
Teacher	3.3	5.3	6.5	9.8	6.2	11.4	8.0	55
Protestant minister	19.8	23.7	15.6	8.8	5.6	6.8	10.4	72
Factory owner, manufacturer, banker, estate owner	1.1	5.3	6.5	10.8	17.4	11.8	10.6	73
Merchant, trader, businessman	12.1	7.9	19.5	14.7	20.5	25.0	19.2	132
Low-ranking civil servant, craftsman, laborer, peasant, innkeeper	28.6	18.4	13.0	15.7	8.7	7.7	13.1	90
Total (in %)[a]	100	100	100	100	100	100	100	
N	91	38	77	102	161	220	689	

Note: Based on Drüll (1986). Coding of biographical data and calculation by authors.
[a]Rounded to nearest 0.1%.

of the professorships shrank by about three-quarters from the period before 1830–1910. The sons of junior civil servants, artisans, peasants, and innkeepers, fared little better. World War I and the economic crisis of the 1920s led to a marked change in German society. By 1932, the proportion of professors' sons among Heidelberg's professors had declined by more than a quarter since the years from 1891 to 1910, and the share accounted for by sons of manufacturers, industrialists, bankers, and landowners had diminished by a third.

The father's social status and occupation, the sociocultural family milieu in which professors had spent their childhood and youth and in which they developed their professional aspirations, and the financial support a family could give its children heavily influenced the son's choice of subjects as well. In Germany as a whole, university students from the lower classes favored theology and disciplines that trained grammar school teachers. Among professors in the arts and humanities (including the social sciences and economics), the sons of Protestant clergy, junior civil servants, and teachers were conspicuously overrepresented, and the sons of manufacturers, industrialists, bankers, landowners, and university professors were the least well represented. Because law graduates were expected to work without pay for 3–10 years before they could expect a salaried administrative post or a position in the legal system (Anderson, 2004, p. 15; McClelland, 1980, p. 43; Turner, 1987, p. 241), this area of study was somewhat risky for upwardly mobile students from the relatively poor classes of society. Law was seen as a standard form of liberal education for wealthy or aristocratic young men. The study of medicine was also relatively expensive and took longer than other subjects. Among the professors of medicine, the sons of doctors, apothecaries, and university professors were overrepresented, and the children of clergymen, judges, public prosecutors, lawyers, and junior civil servants were obviously underrepresented. Among the professors of the natural sciences, the children of manufacturers, industrialists, bankers, and landowners; of skilled tradesmen, laborers, peasants, and innkeepers; and of merchants, traders, and entrepreneurs were disproportionately well represented, with the children of teachers, junior civil servants, and clergy being clearly underrepresented. A direct inheritance of status (cases in which both father and son were university professors) was far more pronounced in medicine and the natural sciences (13.7%) than in the humanities, the social sciences, and economics (9.7%).

The social origin of professors also affected the course of their academic careers. Heidelberg professors whose fathers were senior civil servants completed their doctorates on average at the age of 23.4 years; the children of university professors, 24.0 years; and the children of teachers, junior civil servants, skilled craftsmen, laborers, peasants, and innkeepers, 25.0 years. The sons of university professors completed their *Habilitation* on average at the age of 27.5 years; the children of teachers, 32.2 years. The children of professors also received their first professorial appointment roughly 4 years earlier (aged 32.3 years) than the upwardly mobile graduates from teachers' families (36.6 years). Although these differences are to some extent linked with the dissimilar preferences for and requirements of the disciplines, they are also partly explicable by class-specific disparities in graduates' level of information and ambition, their social networks, and the support from their families.

Religious Denomination of Professors

At Heidelberg University the principle of religious tolerance had been cited as a desirable goal in 1746 and again in 1786 (Weisert, 1974, p. 86). The Religious Declaration of May 9, 1799, officially ended the institution's rigorous confessional orientation (Wolgast, 1986a, p. 85). In 1803 the 13th Organizational Edict expressly declared that the most suitable and competent candidates should be appointed as professors no matter what their religious affiliation. At many universities, however, the conflicts between members of the Reformed Church and Lutherans, and later between Protestants and Catholics, dragged on to the end of the nineteenth century (see Turner, 1987, p. 228). In effect, religious confession remained a criterion for the appointment of professors until the early twentieth century (Dickel, 1961, p. 212).

Up to 1785 Catholics accounted for most of Heidelberg University's newly appointed professors because the rulers at that time were themselves Catholic and because the University of Heidelberg was under strong Jesuit influence. Between 1795 and 1932, though, five and a half times more Protestants than Catholics were appointed. Despite various tolerance edicts, according to which religious confession was no longer to figure in such appointments, the share of Protestant professors climbed from 65.4% in the period before 1830 to 85.8% from 1891 to 1910, whereas that of Catholics fell from 33.6 to 9.3% (Table 4). This conspicuous shift is not explicable merely by the fact that the Catholic theological faculty was moved from Heidelberg to Freiburg in 1807. The proportion of Catholic professors dropped steadily in every decade of the nineteenth century, reaching its nadir between 1891 and 1910.

There are several reasons for the low proportion of Catholic professors, which was also typical of other German universities. First, a higher percentage of Catholics

Table 4 Religious confession of professors (*N*) employed at Heidelberg University

Period of first appointment in Heidelberg	Confession										Conversions from Jewish to Christian	
	Protestant		Catholic		Jewish		Without religious affiliation		Total			
	%	N	%	N	%	N	%	N	%	N	%	N
Before 1830	65.4	70	33.6	36	0.9	1	–	–	100	107	2.8	3
1831–1850	76.3	29	21.1	8	2.6	1	–	–	100	38	2.6	1
1851–1870	75.9	60	15.2	12	8.9	7	–	–	100	79	1.3	1
1871–1890	78.9	86	11.9	13	8.3	9	0.9	1	100	109	5.5	6
1891–1910	85.8	139	9.3	15	2.5	4	2.5	4	100	162	6.8	11
1911–1932	71.8	163	15.4	35	10.1	23	2.6	6	100	227	4.0	9
Total	75.8	547	16.5	119	6.2	45	1.5	11	100	722	4.3	31

Note: Biographical data from Drüll (1986). Coding of biographical data and calculation by authors.

than Protestants lived in rural areas or pursued less urban professions and were generally underrepresented among grammar school and university students. According to Nipperdey (1988, pp. 38–39), Catholics in the second half of the nineteenth century belonged to the premodern, precapitalist, preindustrial social world of the traditional agricultural middle class and were underrepresented in the burgeoning tertiary sector (trade, banking, insurance, and administration). Both the conflict between Catholics and liberals and the struggle between Church and State over the control of the school system in the second half of the nineteenth century also contributed to the frequent hesitation of Protestant-dominated faculties to appoint Catholic applicants despite their excellent qualifications. At the end of the nineteenth century, Catholicism was opened and modernized. Intergenerational social mobility increased after World War I, expanding the share of Catholics among Heidelberg's professors again from 9.3 to 15.4% in the last two periods studied (1891–1910 and 1911–1932) and reducing that of Protestants from 85.8 to 71.8%.

The Jewish presence at Heidelberg University developed quite differently. Although the plan to appoint Spinoza as a professor at Heidelberg University in 1673 failed, Heidelberg was one of the first German universities to accept Jewish students (two from Mannheim in 1724), to graduate Jewish students (1728), and to appoint Jews as professors (Richarz, 1974, pp. 29, 33). In absolute terms, the University of Halle, one of the most modern German universities of the time, drew far more Jews than Heidelberg did (Richarz, 1974, p. 47). The first Jewish professor in our sample of Heidelberg's faculty is Daniel Wilhelm Nebel. He completed his doctorate in Heidelberg in 1758, was appointed as an associate professor of medicine at Heidelberg University in 1766, and became a full professor in 1771. By 1860 two more Jews had become professors at Heidelberg, as had four Jews who had converted to Christianity.

However, the real rise of Jewish professors at Heidelberg University did not begin until after the revolution of 1848. Between 1851 and 1870, 8.9% of Heidelberg's newly appointed professors were Jewish. Traditionally, German Jews were open to emancipation and acculturation and were associated politically with liberalism. According to Wolgast (1989, p. 19), however, German liberalism took a nationalist turn in the Bismarck era and gradually became anti-Semitic. The economic depression of the 1870s and the immigration of Eastern European Jews provided a new platform for anti-Semitism. The decline of Jewish representation to 2.5% of the professors appointed in Heidelberg between 1871 and 1910 strongly suggests that prospects of an appointment as a professor of Jewish academics eroded in the late nineteenth century.

The effect of anti-Semitism is also apparent from the fact that the proportion of professors who had converted from Judaism to Christianity again grew considerably after 1870. In the period from 1871 to 1890 converted Jews accounted for 5.5% of the appointments to professorships at Heidelberg University, reaching 6.8% from 1891 to 1910. Of the professors appointed in Heidelberg between 1911 and 1932, 10.1% were Jewish. More than half (51.1%) of the 45 Jewish professors employed between 1803 and 1932 at Heidelberg University were appointed between 1911 and

Table 5 Social origin and religious confession of professors (N) appointed at Heidelberg University, 1803–1932

Father's occupation	Protestant		Catholic		Jewish		Jewish converts to Christianity		Total	
	%	N	%	N	%	N	%	N	%	N
University professor, scientist, or other kind of scholar	13.3	71	5.8	6	7.3	3	–	–	11.6	80
Senior official, senior civil servant, judge, attorney, lawyer, doctor, apothecary, politician, public prosecutor, commissioned officer	24.9	133	44.2	46	14.6	6	20.0	2	27.1	187
Teacher	9.2	49	5.8	6	–	–	–	–	8.0	55
Protestant minister	12.9	69	1.9	2	2.4	1	–	–	10.4	72
Factory owner, manufacturer, banker, estate owner	10.1	54	8.7	9	19.5	8	20.0	2	10.6	73
Merchant, trader, businessman	18.2	97	8.7	9	48.8	20	60.0	6	19.2	132
Junior civil servant, craftsman, laborer, peasant, innkeeper	11.4	61	25.0	26	7.3	3	–	–	13.1	90
Total	100	534	100	104	100[a]	41	100	10	100	689

Note: Biographical data from Drüll (1986). Coding of biographical data and calculation by the authors.
[a] Rounded to nearest 0.1%.

1932. Including those professors who had converted from Judaism to Christianity, Jews accounted for 10.5% of all professors at Heidelberg University.

When interpreting these numbers, one should bear in mind that, according to the census of 1910, Jews accounted for only 1.2% of the total population of Baden and 0.95% of the population of the Second German Reich (Titze, 1987). Thus the proportion of Jews among Heidelberg's professors was more than ten times that of Jews in the population as a whole (for explanations see Volkov, 1987). By comparison, the proportion of Catholics among Heidelberg's professors corresponded to only a quarter of their share in the German population as a whole. In Prussian universities the percentage of Jewish students in the 1886–1887 academic year was eight times greater than that of Jews in the total population (Richarz, 1974, p. vii). Similar proportions were to be found in other German provinces. In Württemberg there were 14.57 Protestant, 16.10 Catholic, but 101.29 Jewish university students per 10,000 males in the 1909–1910 academic year (Rienhardt, 1918, p. 46).

The religious affiliation of professors was also closely linked to their social background (Table 5). In Heidelberg, most Jewish professors came from relatively well-to-do backgrounds. Slightly more than 68% of them had a father who was a merchant, trader, entrepreneur, manufacturer, industrialist, banker, or landowner. Among Jews who had converted to Christianity, the corresponding figure was even higher (80%). The percentage of Protestant professors from that milieu was much lower; that of Catholics, lower still. A disproportionately high percentage of Protestant professors came from families where the father was a senior civil servant; a merchant, trader, or entrepreneur; a junior civil servant; a university professor; or a clergyman. Catholic professors came mostly from families in which the father was a senior civil servant, judge, public prosecutor, lawyer, doctor, apothecary, soldier, or politician. About one quarter of Heidelberg's Catholic professors had a father who was junior civil servant, skilled craftsman, peasant, or innkeeper. The direct inheritance of status (either having or having had a university professor for a father) was the most common among the Protestant professors and the least common among the Catholic ones. Direct inheritance of status was relatively low among Jewish professors.

Academic Careers, the Appeal of Places, and the Importance of Spatial Relations

Theoretical Aspects

Two questions of particular interest in this chapter have figured in discussions about geographies of science (e.g., Livingstone, 1995, 2003) and sociologies of science (Ophir & Shapin, 1991; Shapin, 1998). One is whether the location of scientific endeavor affects the substance of science and why scientific theory is occasionally modified or transformed when it travels from one culture to the next. The other is whether the sites where experiments are conducted, the places where new knowledge is generated, or the localities where investigation is carried out have any bearing on whether a scientific claim is accepted or rejected.

There are at least two reasons for the reciprocal relationship between the prestige of places and that of individual scholars. First, numerous biographies of scholars confirm that the places at which training, research, and the evaluation of research findings take place have definite significance for learning processes, the accumulation of experience, the perception of problems, the formulation of research topics, the acquisition of disciplinary competence, and integration into key networks. The quality of scientific work depends on the training a scholar has received, the equipment and networks available to him or her, the supervisory role models that person has encountered, and the experiences inspiring or discouraging his or her research. In specific career periods, the places at which a scholar's life path becomes bound up with those of other scholars are critical. Different places afford different degrees of support by supervisors. The more distinguished a scholar is in comparison to colleagues, the more freely he may showcase his disciples (Caplow & McGee, 1958, p. 72). In addition, a university's location in a specific environment can raise research questions overlooked at other places and can facilitate the development of competence in special domains, especially in disciplines where fieldwork plays a salient role.

Second, the human tendency and necessity to reduce complexity means that the esteem of hundreds of scholars who have contributed to a university's academic standing over a given period is transferred to the institution or place at which they have researched and taught. By the same token, the prestige an institution has gained over a long period is projected onto individual scholars appointed there. The general assumption is that the particular quality of teaching and research offered at certain universities—independent of their continuously fluctuating academic staff—is commensurate to their resources, reputation, and appointment policy. The public may therefore trust in the academic eminence of Harvard, Cambridge, or Heidelberg without knowing any of the academics working there. The higher a department ranks in its discipline, the better the reputation of its individual members. Both graduation from renowned departments and appointments to respected faculties build symbolic capital and can be used as reference points to reduce uncertainty about the scientific potential of scholars. If two or more illustrious faculties have acknowledged a scholar and have selected him or her from a host of applicants, uncertainty about the candidate's scientific potential diminishes more than if that person's academic reputation has been recognized only by the institution at which he or she earned a doctorate or completed the *Habilitation*.

Why do many academic cultures expect their scholars to be spatially mobile? First, the transmission of scientific knowledge between different environments is far more convoluted and time-consuming than is widely assumed. The spatial diffusion of ideas, instruments, theories, and inventions from the place of origin to radically dissimilar environments can be accomplished best by the circulation of scholars who have generated them or who are proficient in them (see Livingstone, 2003, pp. 140, 177–178; Meusburger, 2008, pp. 69–74). Second, faculties want young scholars to gather experience and cope with challenges in a variety of environments. Scholars should demonstrate that their scientific achievements are willingly accepted by different faculties. Such changes of location do not have to be permanent. After all, circular mobility that temporarily takes an academic to other universities and then back home can initiate just as many new stimuli and learning processes and just as much inspiration as a permanent move (see Jöns, 2003). In many countries academic mobility is expected in order to prevent faculty in-breeding and disciplinary constriction. It is assumed that a faculty can achieve a high academic standard only if it does not exhibit partiality to its own graduates in the appointment of academic staff but instead selects the best available scholars from outside, using stringent, transparent, and objective procedures to do so. Measures inducing academic mobility and hindering academic in-breeding are partly a reaction against the "aggrandizement effect" (Caplow & McGee, 1958, p. 45)—the tendency to overrate one's own department, inflate the ratings of one's own research group vis-à-vis competing groups, and disparage other departments in the same disciplines. Caplow and McGee (1958) found that raters overestimated the regard for their own organization eight times more frequently than they underestimated it (p. 105). Some of this phenomenon's effects can be avoided if a department's academic offspring must move on and if qualified scholars are brought in from other universities.

Structures of Academic Faculties in Nineteenth-Century German Universities

Without background knowledge about differences between full professors, associate professorships, and unsalaried university lecturers, one will find it difficult to understand many trends and structures of the German university system. Any inquiry into the structure of the academic faculties at German universities in the nineteenth century must deal with the problem that the same occupational title can cover exceedingly diverse activities, rights, and salaries. In the university system of German-speaking countries, the only plainly definable, homogeneous group of academics with salaried posts was that of the full professors, who were entitled to exercise university self-government.

Associate professors and university lecturers formed two remarkably heterogeneous groups not clearly definable by age or responsibilities. A lecturer could be the same age as an associate professor. Some associate professors and lecturers could confidently expect appointments to full professorships, were closely associated with the university, and became important pioneers of their discipline. Other academics in these two groups were employed outside the university (e.g., hospitals) and attained lecturer status at the peak of their professional careers. In some states (e.g., Prussia) the associate professors received a fixed salary and a specific teaching post, in others they did not (for further details see Daude, 1896; Eulenburg, 1908; Nauck, 1956; Schmeiser, 1994). This lack of uniformity and precision in the definition of scholars who were not full professors complicates the empirical analysis of academic careers. A sharp distinction between full and associate professors did exist, however, when it came to decision-making in faculties and departments. For this reason the tables in this study discriminate between those two groups only.

Nonetheless, it is useful to discriminate between two types of associate professors. The first type was usually recruited from another university in order to establish a new discipline or specialized research field. Having a narrower specialization than full professors did, some of these associate professors became leading pioneers in the development of new disciplines. They received a regular salary from the state budget, and many of them were promoted to the rank of full professor after a few years if their field developed successfully. One example of this type of scholar was the geographer Alfred Hettner (1859–1941), who became an associate professor at Heidelberg University in 1899 and a full professor in 1905. The physicist Helmholtz was called from Berlin to the University of Königsberg as an associate professor of physiology and pathology in 1849 and became a full professor in 1852. The second type of associate professor was not on the state payroll and received only lecture fees paid by the students (*Kollegiengeld*). It was a kind of waiting post, and its incumbent functioned mainly as a kind of adjunct professor helping out with the teaching load. In practice, service as an unsalaried university lecturer became an almost obligatory stage in which teachers developed their academic credentials. These lecturers could receive only the fees paid by students who attended their lectures. By the 1880s, however, introduction of the institute, seminar, and teaching laboratory, where students received instruction from associate professors and unsalaried lecturers in small groups and prepared original projects that could eventually become theses, created a new demand for assistants and associate professors who would provide supplementary teaching and supervise laboratory work. The system of untenured and low-paid university lecturers tended to preclude academic careers for men without family financial support. But it did create an intensely competitive academic labor market, allow development of new specialties in disciplines (thereby bypassing the monopolies of full professors), and create a pool of talent from which future professors could be drawn (for details see Anderson, 2004, pp. 60, 153).

The positions of unsalaried lecturer and associate professor enabled universities in German-speaking countries to increase their specialization and the number of seminars and institutes without permanently committing themselves to new subdisciplines, incurring the expense of new chairs, or diluting the power of full professors in university matters (McClelland, 1980, p. 166).

Careers, Regional Provenance, and the Mobility of Professors

In this section we examine indicators significant at an aggregate level for structural changes in university staffing and changes in academic reputation.

Age Upon Completion of the Habilitation

Rising academic standards confronted young scholars with ever greater barriers to earning a doctorate and the postdoctoral degree, the *Habilitation*. Professors of Heidelberg University who began their careers in the period between 1831 and 1850 completed their doctorates on average at the age of 23.2 years. Thereafter, the average age at which doctorates were completed rose in most faculties, peaking at 25.4 years among those professors who began their university careers between 1911 and 1932. But much more noteworthy is the age at which the *Habilitation* was received. At Heidelberg University, the first regulations on the *Habilitation* date from 1803 (Nauck, 1956, p. 25). Only 30.2% of those professors who had their first appointment in Heidelberg before 1830 held that degree. But this figure exceeded 90% by 1890. The most impressive rates were achieved in the faculty of science (93.4%), followed by the faculties of medicine (87.5%), law (83.1%), the arts (80.0%), and theology (66.7%).

After 1850 the requirements for *Habilitationsschriften*, or postdoctoral dissertations, became far more demanding, a change that lengthened the time necessary for this qualification (Table 6). Among those Heidelberg professors who began their careers before 1830, the lapse of time between the completion of the doctoral dissertation and the *Habilitationschrift* was only 1.5 years. For those who received their first appointment between 1871 and 1890, the time needed to finish the *Habilitationsschrift* had extended to 4.5 years, stretching to 6.8 years for those beginning their careers between 1911 and 1932.

The average age of Heidelberg's professors on completion of their *Habilitation* went up correspondingly by 6 years during the period studied here. The professors who began their first appointment before 1830 finished their *Habilitation* on average

Table 6 Average age upon completion of the *Habilitation* (postdoctorate) and upon first appointment as a professor at any university (1803–1932)

Period of first appointment	Average age upon completion of postdoctorate	Interval between completion of doctorate and postdoctorate in years			Average age upon the first appointment as professor	
		Full professors	Associate professors etc.	All scholars	Full professors	Associate professors
Before 1830	25.73	0.33	1.63	1.46	34.73	31.00
1831–1850	25.98	1.25	2.32	2.21	37.00	31.00
1851–1870	28.26	3.50	3.81	3.79	33.40	33.85
1871–1890	28.26	4.47	4.50	4.49	31.11	33.00
1891–1910	29.84	4.40	5.32	5.27	34.42	34.15
1911–1932	31.84	5.29	6.83	6.77	34.88	38.01
Total	29.56	3.67	5.03	4.92	33.93	34.46

Note: Biographical data from Drüll (1986). Coding of biographical data and calculation by the authors. Professors without a *Habilitation* are not represented.

at the age of 25.7 years. This average age went up to 28.3 years in the years from 1851 to 1870 and to 31.8 years from 1911 to 1932. In the period from 1803 to 1932 as a whole, professors of law were on average the youngest to complete their *Habilitation* (28.5 years), trailed by theologians (29.8 years) and natural scientists (29.8 years). Professors of the medical faculty were the oldest in this sense (30.3 years).

The age upon completion of the *Habilitation* has proven to be a relatively reliable early indicator for a successful academic career. Those academics who became full professors upon their first appointment (usually after completion of their *Habilitation*) were quite a bit younger when they finished that work than were their fellow professors whose first appointment was to an associate professorship (Table 7).

The marked upward trend in the average age at completion of the *Habilitation* suggests a corresponding shift in the average age of the candidates at their first professorial appointment. However, this expected trend in the age of newly appointed professors is apparent only for those academics whose first appointment was as an associate professor. The average age of professors whose first appointment was as a full professor (at any university) changed little over 130 years, varying between 34.7 and 34.9 years with a minimum of 31.1 years from 1871 to 1890. Among associate professors, however, the average age at their first appointment soared from 31.0 to 38.0 years, primarily because the number of applicants in that pool mounted more rapidly than the number of available professorships. Whereas the recruitment of full professors could tap into a steadily accruing body of excellent applicants—an advantage that greatly contributed to the swift advance of German universities—the supply of associate professorships could not keep pace with the number of young academics who had completed a *Habilitation*. The time that professors in the latter group had to wait for promotion therefore became quite long.

Table 7 Age distribution of Heidelberg professors upon completion of their *Habilitation* (postdoctoral degree)

	Percentage of professors in age group upon conferral of the *Habilitation*							
	Full professors		Associate professors		Others		Total	
Age group	%	N	%	N	%	N	%	N
25 years and younger	34.4	55	15.9	66	21.9	7	21.1	128
26–30 years	46.9	75	44.4	184	68.8	22	46.4	281
31–35 years	15.0	24	27.1	112	3.1	1	22.6	137
36–40 years	3.1	5	8.0	33	–	–	6.3	38
Older than 40 years	0.6	1	4.6	19	6.2	2	3.6	22
Total	100	160	100	414	100	32	100	606

Note: Biographical data from Drüll (1986). Coding of biographical data and calculation by authors. Professors without a *Habilitation* are not represented.

Number of Appointments Offered to Heidelberg's Professors from Other Universities

Because it is easier to be appointed as professor at a lower-ranking university than at a highly regarded one, quite a few scholars start their professorial careers at the academic periphery of their disciplines—a new, small, or less well-equipped university, where professors' salaries are lower and resources smaller. After promotion to full professor, academics could no longer enhance their rank—only their institutional prestige, salary, and research infrastructure—by moving from a university of modest reputation to one at the forefront of their discipline. In the university system of German-speaking countries, large gains in a professor's salary or in a department's budget, research infrastructure, and size of staff can be achieved predominantly by negotiating the conditions for acceptance of a full professorship at *another* university. A professor who has received a call to another university will first negotiate with the prospective university about the department's budget, research infrastructure, and size of staff. If the professor's present university wants to keep that person, the professor can also negotiate for a bigger budget and staff at his or her home university and then choose the superior deal. Each offer a professor receives therefore affords an opportunity to improve his or her personal and departmental research conditions. And the higher the number of outside offers received, the stronger his or her negotiating position, the more lucrative the salary, and the better the research infrastructure of his or her group. This mobility affects the standing of both the individual scholar and the institutions involved. It can be assumed that established academics will change their university only when the new location is much more attractive than the previous one or if it promises a clear improvement in research conditions. Because each decision of a full professor to accept or decline a call to another university resembles a personal value judgment about his or her subjectively perceived situation at the two institutions, a university's acclaim and magnetism is deducible also from its capacity to lure eminent academics from other universities.

The number of appointments a professor receives in his or her career is one of the most telling indicators of the scientific reputation that this person enjoys among the academic gatekeepers of the given discipline. However, the number of full professorships to be occupied and the number of academics applying for them fluctuate over time. Though there are exceptions to the rule (Immanuel Kant, one of the most celebrated of all German scholars, never left the University of Königsberg), this indicator is quite meaningful on aggregate. As shown in Table 8, the average number of appointments offered to professors of Heidelberg University over their academic careers went up noticeably from the period before 1830 (1.85) to 1932 (3.37). The average number of appointments accepted by Heidelberg's full professors more than doubled (1.23–2.64).

The growing share of experienced professors who had already completed several stages of their academic career before appointment to full professorships at Heidelberg University demonstrates the appreciation in that institution's pull and scientific renown between 1803 and 1932 (Table 9). The percentage of professors or

Table 8 Average number of appointments (*M*) offered to full professors (*N*) employed at Heidelberg University, 1803–1932

Period of first appointment in Heidelberg	Accepted		Refused		Total	
	M	*N*	*M*	*N*	*M*	*N*
Before 1830	1.23	84	0.62	84	1.85	84
1831–1850	1.44	25	0.32	25	1.76	25
1851–1870	1.91	34	0.56	34	2.47	34
1871–1890	2.29	52	0.58	52	2.87	52
1891–1910	2.44	63	0.70	63	3.14	63
1911–1932	2.64	78	0.73	78	3.37	78
Total	2.03	336	0.63	336	2.66	336

Note: Biographical data from Drüll (1986). Coding of biographical data and calculation by authors.

unsalaried university lecturers who became full professors at Heidelberg University as their first appointment after completing their *Habilitation* shrank substantially from a high of 19% during the period before 1830 to 4.8% in the years from 1891 to 1910 and to a mere 1.3% by the two decades from 1911 to 1932. Over the same period, the percentage of the professors who did not attain a chair at Heidelberg University until their third professorship nearly doubled. The share of Heidelberg's professors who had to wait even longer in their careers for a chair at Heidelberg University more than quintupled from the period before 1830 to the years between 1891 and 1911, receding only modestly in the final period under review. The pattern varied from discipline to discipline, though. In the Faculties of Theology and the Arts and Humanities, a distinctly higher proportion of professors were appointed to full professorships as the first step in their career (13.6 and 12.8%, respectively) than

Table 9 Number of appointments that lead to a full professorship at Heidelberg University

Appointment in Heidelberg	1st appointment		2nd appointment		3rd appointment		4th appointment		5th–9th appointment		Total	
	%	*N*	%	*N*	%	*N*	%	*N*	%	*N*	%	*N*
Before 1830	19.0	16	52.4	44	22.6	19	4.8	4	1.2	1	100	84
1830–1850	8.0	2	60.0	15	24.0	6	4.0	1	4.0	1	100	25
1851–1870	8.8	3	41.2	14	32.4	11	14.7	5	2.9	1	100	34
1871–1890	13.5	7	26.9	14	28.8	15	21.2	11	9.6	5	100	52
1891–1910	4.8	3	20.6	13	42.9	27	20.6	13	11.1	7	100	63
1911–1932	1.3	1	26.9	21	42.3	33	14.1	11	15.4	12	100	78
Total	9.5	32	36.0	121	33.0	111	13.4	45	8.0	27	100[a]	336

Note: Eight professors (2.4%) had more than one appointment in Heidelberg. Biographical data from Drüll (1986). Coding of biographical data and calculation by authors.
[a]Rounded to nearest 0.1%.

was the case in the Faculties of Law (7.0%), Medicine (6.0%), and Natural Sciences (3.7%). In the Natural Sciences, 40.7% of all full professors were appointed to their posts in Heidelberg only upon their fourth career move or even later. The corresponding figure for the Faculty of Arts and Humanities was (23.2%); Law (19.3%); Medicine (18.0%); and Theology (13.7%).

Relations Between Academic Standards of Universities and Catchment Areas of Scholars

Academic mobility and regional provenance of professors and students developed more or less concurrently with the ups and downs of scientific reputation, at least in peacetime. Whenever the appointment and promotion of professors did not abide by meritocratic principles but rather guild-like procedures favoring nepotism and seniority (as during eighteenth century), local candidates were preferred and nation-wide searches for the best candidates tended to be an exception. In periods of intellectual stagnation, most scholars did not have to leave their universities to get promoted. They received internal, or in-house, promotions to professorships when such posts became vacant at their universities (McClelland, 1980, p. 81). It was mainly Göttingen's new recruitment policy of not permitting internal promotion (for details see McClelland, pp. 40–41) that restored a measure of mobility to the German system. Göttingen was determined to draw famous and excellent men from all over Germany and from foreign countries. This resolve was a significant reaction against the regionalist spirit of the German *Landesuniversität* that dominated in the eighteenth century (for details see Anderson, 2004, pp. 24–25). The outstanding offers from Göttingen became almost irresistible, won the best scholars available, created faculties of international repute, and thus expanded the catchment area of the professors—that is, the geographical zone from which they came. Universities aiming to be competitive had to emulate Göttingen's example.

In Germany, however, the training of professional elites has always been distributed across more universities than was the case in Britain. Moreover, the ranking of universities was not as stable in Germany as it was in Britain, where Oxford and Cambridge had long monopolized the education of the British elites and had shaped the value system of the upper classes. The ranking of German universities could change within a relatively short period of time. Whereas Göttingen, Halle, Jena, and Erlangen topped the list of German universities in the eighteenth century, Berlin University became a model for many other universities in Europe after 1810. After 1850 Heidelberg and Munich joined the vanguard of German universities. The rivalries among German universities for the best scholars and the competition among candidates for the most coveted chairs not only created a powerful stimulus to intellectual progress and served as the main source of the research vitality characterizing German universities (Anderson, 2004, p. 61), it also fostered spatial mobility. As a university's importance and esteem increased, so did the objective criteria of its recruitment procedures, external demand for its expertise, the barriers against in-breeding, the frequency of nationwide searches for the best candidates, the status attached to the mobility of its professors, and the geographic scope and differentiation of the areas from which its professors were appointed.

Professors' Places of Birth

At first sight, regional provenance has little to do with academic excellence. No appointment committee will normally be interested in the birthplaces of scholars. However, moving beyond the microlevel of individual scholars and interpreting spatial patterns that emerge from a relatively sizeable population of professors can produce new insights that are otherwise obscured. The spatial distribution pattern of the birthplaces of a university's professors and the dynamic changes of those patterns over time can be a relatively meaningful indicator of various things. It can reveal a great deal about recruitment policies, the development of disciplines, the role of language barriers and political conflicts, the permeability of political borders, and so on. A wide distribution of birthplaces implies a broad range of formative learning environments, challenges, cultural experiences, value systems, and attitudes that, together with other influences, might help prevent academic parochialism. Just imagine how North American behavioral and social sciences would have developed without the influx of thousands of European scholars since the 1930s. Because historical creativity is very rare (for details see Meusburger, 2009), it is most unlikely that multitudes of creative scholars would be born and raised in a small catchment area.

Of the 722 professors who taught at Heidelberg University between 1803 and 1932, 14.6% of the full professors and 26.1% of the associate professors were born in the territory of Baden or Württemberg. However, the proportion of professors born in Baden or Württemberg declined continually as Heidelberg University gained in academic prestige (see Fig. 1). Of the 145 professors who were first appointed to Heidelberg before 1850, 11.7% were born in the city of Heidelberg, 33.1% in the territory of the states of Baden or Württemberg, and 66.9% in 79 other places outside those two states (see Fig. 1). Of the 188 professors who were first appointed to Heidelberg between 1851 and 1890, only 2.7% were born in Heidelberg and only 13.8% in Baden or Württemberg. Just over 86% were born in 108 other places of other states. With reference to today's national boundaries, 144 professors of Heidelberg University (19.9%) were born outside the Federal Republic of Germany, 54 of them in modern Poland (particularly Silesia), 17 in Switzerland, 14 in Russia (especially in former East Prussia), 11 each in Austria and France, 10 in the Czech Republic, 6 each in Ukraine and Romania (Transylvania), 2 each in The Netherlands, Italy, Estonia, Latvia, and Slovenia, and 1 each in Lithuania, Slovakia, the United States, and Mexico. For those readers not familiar with all the place names appearing on the map, it might be more informative to learn that the proportion of Heidelberg's professors born within a 100 km (62 mile) radius of Heidelberg University ebbed from 43.4% (first appointment before 1830) and 38.5%

Fig. 1 Birthplaces of Heidelberg's full professors, 1803–1932

(first appointment 1830–1870) to a mere 23.2% (first appointment 1871–1910). It edged up 30.0% in the period from 1911 to 1932 because of the swelling proportion of associate professors appointed in those years.

Places Where Heidelberg's Professors Earned Their Doctorate and Habilitation

Overall, the 722 professors of Heidelberg University completed their doctorates at 40 university locations in what is now Germany, Austria, Switzerland, France, The Netherlands, the Czech Republic, Estonia, Russia, Poland, Hungary, and Ukraine.

As is to be expected, the places at which full professors completed their doctorates were more widely distributed than those of the associate professors. Just over 44.4% of the associate professors completed their doctorates in Heidelberg, followed by Berlin (8.4%), Munich (4.3%), Göttingen (3.9%), Bonn and Freiburg (3.6% each), Jena and Leipzig (3.2% each), Strasbourg (2.3%), Giessen (2.0%), and Vienna (1.6%). By comparison, only 20.8% of the full professors were awarded their doctorate at Heidelberg University, whereas 11.6% were granted their doctorates in Berlin, 7.7% in Leipzig, 5.7% in Göttingen, 4.5% in Bonn, 4.2% each in Munich and Strasbourg, 3.9% in Jena, and 2.7% each in Tübingen and Vienna (see Fig. 2).

The 722 professors completed their *Habilitation* or were appointed as unsalaried university lecturers at a total of 39 university locations in today's Germany, Austria, Switzerland, Italy, France, Poland, the Czech Republic, Estonia, Russia,

Fig. 2 Places where Heidelberg's full professors completed their doctorates, 1803–1932

and Ukraine (see Fig. 3). The differences between full professors and associate professors are even more pronounced when it comes to the *Habilitation*
and appointments as an unsalaried university lecturer than in case of doctorates. Of Heidelberg University's associate professors, 77.1% had completed their
Habilitation in Heidelberg, 1.6% in Jena, 1.4% in Göttingen, and 0.9% each in
Berlin, Kiel, and Strasbourg. The remaining 17.2% of Heidelberg's associate professors completed their *Habilitation* at 20 other universities, and 9.3% had no
Habilitation. Of the full professors, only 19.3% completed their *Habilitation* in
Heidelberg, 6.8% in Berlin, 6.3% in Leipzig, 5.1% in Göttingen, 4.2% in Bonn,
3.3% in Marburg, and 3.0% each in Halle and Munich. The remaining 49%
were awarded their *Habilitation* at 25 separate universities, and 24.4% had no
Habilitation. However, the differences between full and associate professors as
noted in Figs. 2 and 3 should not be mistaken for value judgments indicating disparities in scientific excellence. They are, rather, the result of the career stages involved.

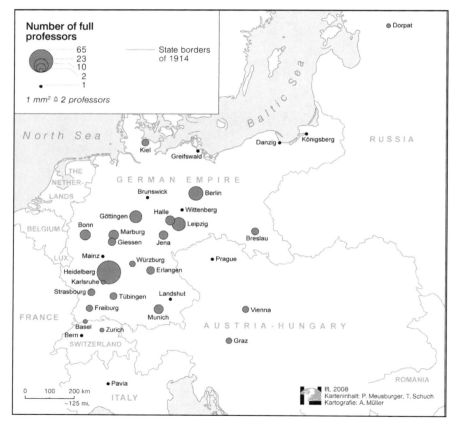

Fig. 3 Universities at which Heidelberg's full professors received their *Habilitation* (postdoctoral
degree), 1803–1932

In the course of the nineteenth century, the number of places at which Heidelberg professors completed their *Habilitation* grew continually. The professors appointed before 1851 had been awarded their *Habilitation* at 22 universities, whereas those appointed between 1891 and 1932 had completed that work at 30 universities.

Places from Which Academics Were Appointed as Professors in Heidelberg

In the German university system today, warnings against faculty in-breeding are mainly directed at internal promotions. An internal promotion is defined in this context as the appointment of a scholar to the rank of professor if that person has earned the *Habilitation* within the same faculty as the one to which he or she accedes to that position without first receiving a call to another university. The stigma of faculty inbreeding does not apply to the places where the doctorates have been earned nor to those scholars who completed their *Habilitation* at Heidelberg University, became a professor at one or more other universities, and returned to Heidelberg. Neither does it apply to persons who received, but declined, an appointment as professor at another university immediately after receiving the *Habilitation* at Heidelberg. In other words, more than one faculty should confirm that the scientific achievements, competence, and other merits of a scholar warrant that person's appointment to a professorship. Internal promotions are frequently subject to the suspicion that a faculty's or department's own member has had unjustified preferential treatment and that academic standards have slipped, creating inappropriate "schools" and narrowing scientific perspectives. In German-speaking countries economic considerations, too, discourage internal appointments. A department or faculty can expect a greater improvement of its scientific infrastructure if it acquires its professors from another university's pool of graduates than from its own. Scholars promoted in-house have a weak negotiating position, so the financial and personnel resources they can secure for the development of their departments are less likely to be as bountiful as those obtainable by scientists appointed from other universities.

A different set of standards was applied to associate professors chiefly because a significant proportion of them did not have a salaried position. Of the 650 associate professors studied by Eulenburg (1908) in German-speaking countries, 25.4% received no salary at all in 1907. In Prussia 19.1% were unsalaried. At Heidelberg University 36.1% of the associate professors received no salary. With regard to the average salaries of associate professors in 1907, Heidelberg University ranked twentieth of 21 universities studied by Eulenburg (pp. 129, 134–135). As long as associate professors received little or no salary, they had no incentive to move to another university for such a post. Another reason for the difference between the standards pertaining to them and those to full professors was that, for many decades, a varying proportion of associate professors were not officially part of the faculty and did not participate in faculty decisions.

Between 1803 and 1932 only 16.3% of Heidelberg University's full professors were internal appointments as opposed to 79.2% of the associate professors. To put it differently, of all the internal appointments at the University of Heidelberg from 1803 to 1932, only 12.6% were full professors and 87.4% were associate professors. Among associate professors, internal appointments rose continually from 55.5% before 1830, peaking at 92.2% between 1851 and 1870. The percentage then dropped to 79.5% between 1911 and 1932. In the individual case, of course, an internal or external appointment says nothing about an academic's scientific qualities or originality. There are numerous examples of an internally appointed academic producing superior scientific achievements and later receiving many offers from other universities. Internal appointments are disadvantageous under only two

Fig. 4 Places where Heidelberg's full professors had been employed before accepting their positions at Heidelberg University, 1803–1932

circumstances: (a) if they exceed a certain proportion of overall appointments in a faculty and (b) if a major share of internally appointed professors receive no other offer from another university until the end of their career. For 38.2% of Heidelberg's associate professors, the appointment in Heidelberg was just the beginning of their academic career or a kind of stand-by position until being called to a chair at another university. For 29.9% of Heidelberg's full professors, Heidelberg University was not their final career station. They left Heidelberg and accepted appointments as full professors at other universities.

The "geographical biographies" (Livingstone, 2003, p. 182) of the appointed full professors and the spatial extent of catchment areas are good indicators by which to distinguish periods of academic openness from periods of parochial university politics. The growing scientific prestige of Heidelberg University led to a marked expansion of its catchment areas and to a decrease in the proportion of in-state professors. The full professors appointed at Heidelberg University from 1803 to 1932 were recruited from 45 universities across 10 states (according to today's borders). Before these scholars were appointed as full professors in Heidelberg, 33.0% of them had already been employed in Heidelberg as unsalaried university lecturers or associate professors. After a period as professors in Heidelberg, they then accepted appointments at one or more other universities and later resumed their full professorships in Heidelberg. We note that many full professors were recruited from leading universities of their day (Fig. 4), with 5.1% of them coming from Switzerland and 1.5% from Austria.

Summary

This chapter has touched on three key topics. First, we discussed various interconnected factors that operated on different scales and enhanced the scientific reputation of Heidelberg University between 1803 and 1932. We have shown how the local concurrence of science-friendly policy, adequate financial resources, uncommon freedom of thought, intellectually inspiring networks, and high standards for appointing and promoting scholars created a specific and unique milieu that attracted outstanding academics and promoted research and creative discourses in an unusual way. This analysis has emphasized that places, environments, and spatial relations play a pivotal role in academic careers, the generation of scientific knowledge, and the legitimization of creative ideas.

Second, this chapter has shown that a university's gradual development from mediocrity to excellence not only raised the academic demands on scholars (as reflected by their career steps) but also enlarged and diversified the catchment areas they were recruited from and expanded the proportion of highly mobile scholars. High standards of recruitment increasingly drew renowned scholars who had already reached the peak of their career and had held chairs at a number of other universities before accepting the call to Heidelberg. The average number of chairs offered to Heidelberg's professors during their career almost doubled during the period under review.

Third, we have examined the religious affiliation and social background of the professors in an endeavor to show how universities function as a mirror or sounding board of societal attitudes and changes. The upheavals of the Counterreformation, the territorial reorganization by Napoleon in 1802, the revolution of 1848, and various other historical events demonstrated how fragile the achievements of the human spirit are (McClelland, 1980, p. 2), how vulnerable an academic system can be, and how much external political influences can affect a university's scientific quality.

Notes

1. By comparison, the statistics reported in the *Stifterverband für die Deutsche Wissenschaft* (December 15, 2008) show that state and business expenditure on research and development in Germany accounted for 2.54% of GDP in 2007. Over the same period, the United States spent 2.61% of its GDP on research and development. The corresponding figure for Japan was 3.39%.
2. Between 1811 and 1914 more than 2,400 students of Russian nationality studied at Heidelberg University.
3. Heidelberg University began admitting women regularly in the academic year 1899–1900.
4. In the first decades of the nineteenth century, the *Habilitation* became an essential qualification for an academic career in the German-speaking states. From then on scholars seeking appointments as professors were expected to publish a *Habilitationsschrift*, a second major piece of research, but one on a topic other than that of the doctoral dissertation.
5. The Board of the 13 kept its name after a fourteenth member was added later.

References

Allert, T. (1995). Max und Marianne Weber. Die Gefährtenehe [Max and Marianne Weber. Companions in marriage]. In H. Treiber & K. Sauerland (Eds.), *Heidelberg im Schnittpunkt intellektueller Kreise. Zur Topographie der "geistigen Geselligkeit" eines "Weltdorfes"* (pp. 210–241). Opladen Germany: Westdeutscher Verlag.

Anderson, R. D. (2004). *European universities from enlightenment to 1914.* Oxford, England: Oxford University Press.

Best, H. (1981). Quantifizierende historische Sozialforschung in der Bundesrepublik Deutschland. Ein Überblick [Quantitative historical social research in the Federal Republic of Germany; An overview]. *Geschichte in Köln, 9,* 121–161. Reprint 2008 in *Historical Social Research* (Supplement No. 20), 49–73.

Breuer, S. (1995). Das Syndikat der Seelen. Stefan George und sein Kreis [The syndicate of souls: Stefan George and his circle]. In H. Treiber & K. Sauerland (Eds.), *Heidelberg im Schnittpunkt intellektueller Kreise. Zur Topographie der "geistigen Geselligkeit" eines "Weltdorfes"* (pp. 328–375). Opladen, Germany: Westdeutscher Verlag.

Busch, A. (1959). *Die Geschichte des Privatdozenten. Eine soziologische Studie zur großbetrieblichen Entwicklung der deutschen Universitäten* [The history of the unsalaried university lecturer: A sociological study on the development of German universities into large organizations] (Göttinger Abhandlungen zur Soziologie und ihrer Grenzgebiete, Monograph No. 5). Stuttgart: Enke.

Caplow, T., & McGee, R. J. (1958). *The academic market place.* New York: Basic Books.

Conrad, J. (1884). *Das Universitätsstudium in Deutschland während der letzten 50 Jahre. Statistische Untersuchungen unter besonderer Berücksichtigung Preußens* [University study in Germany during the last 50 years: Statistical analyses with special consideration of Prussia]. In J. Conrad (Series Ed.), *Sammlung nationalökonomischer und statistischer Abhandlungen des staatswissenschaftlichen Seminars zu Halle a.d.S.: Vol. 3, pt. 2.* Jena: G. Fischer.

Daude, P. (1896). *Die Rechtsverhältnisse der Privatdozenten. Zusammenstellung der an den Universitäten Deutschlands und Österreichs sowie an den deutschsprachigen Universitäten der Schweiz über die rechtliche Stellung der Privatdozenten erlassenen Bestimmungen* [The legal situation of unsalaried university lecturers. Compilation of regulations on the legal situation of unsalaried university lecturers at the universities of Germany and Austria and the German-speaking universities of Switzerland]. Berlin: Becker.

Dickel, G. (1961). Die Heidelberger Juristische Fakultät. Stufen und Wandlungen ihrer Entwicklung [The law faculty of Heidelberg University: Steps and changes of its development]. *Ruperto Carola* [Special issue 1961, H. Hinz (Ed.), *Aus der Geschichte der Universität Heidelberg und ihrer Fakultäten, 1386–1961*], 163–234.

Drüll, D. (1986). *Heidelberger Gelehrtenlexikon 1803–1932* [Encyclopedia of professors at Heidelberg University, 1803–1932]. Berlin: Springer.

Essen, G. von (1995). Max Weber und die Kunst der Geselligkeit [Max Weber and the art of conviviality]. In H. Treiber & K. Sauerland (Eds.), *Heidelberg im Schnittpunkt intellektueller Kreise. Zur Topographie der "geistigen Geselligkeit" eines "Weltdorfes"* (pp. 462–484). Opladen, Germany: Westdeutscher Verlag.

Eulenburg, F. (1904). *Die Frequenz der deutschen Universitäten von ihrer Gründung bis zur Gegenwart* [Attendance at German universities from their creation until the present]. Abhandlungen der königlich sächsischen Gesellschaft der Wissenschaften zu Leipzig, Vol. 24, No. 2. Leipzig & Berlin: Teubner.

Eulenburg, F. (1908). *Der Akademische Nachwuchs. Eine Untersuchung über die Lage und die Aufgaben der Extraordinarien und Privatdozenten* [Young academics: A study on the situation and tasks of associate professors and unsalaried university lecturers]. Leipzig, Germany: Teubner.

Glockner, H. (1969). *Heidelberger Bilderbuch. Erinnerungen von Hermann Glockner* [Storybook of Heidelberg: Memories of Hermann Glockner]. Bonn, Germany: Bouvier.

Jarausch, K. H. (1984). *Deutsche Studenten. 1800–1970* [German university students, 1800–1970]. Frankfurt am Main, Germany: Suhrkamp.

Jellinek, C. (1970). Georg Jellinek. Ein Lebensbild, entworfen von seiner Witwe Camilla Jellinek [Georg Jellinek: A biography outlined by his widow Camilla Jellinek]. In G. Jellinek (Ed.), *Ausgewählte Schriften* (Vol. 1, pp. 7–140). Aalen, Germany: Scientia Verlag.

Jöns, H. (2003). *Grenzüberschreitende Mobilität und Kooperation in den Wissenschaften. Deutschlandaufenthalte US-amerikanischer Humboldt-Forschungspreisträger aus einer erweiterten Akteursnetzwerkperspektive* [Cross-boundary mobility and cooperation in the sciences: U.S. Humboldt Research Award winners in Germany from an expanded actor-network perspective] (Heidelberger Geographische Arbeiten No. 116). Heidelberg, Germany: Selbstverlag des Geographischen Instituts.

Karádi, É. (1995). Emil Lask in Heidelberg oder Philosophie als Beruf [Emil Lask in Heidelberg, or philosophy as a calling]. In H. Treiber & K. Sauerland (Eds.), *Heidelberg im Schnittpunkt intellektueller Kreise. Zur Topographie der "geistigen Geselligkeit 'eines' Weltdorfes"* (pp. 378–399). Opladen, Germany: Westdeutscher Verlag.

Koenigsberger, L. (1919). *Mein Leben* [My life]. Heidelberg, Germany: Carl Winter.

Kolk, R. (1995). Das schöne Leben. Stefan George und sein Kreis in Heidelberg [Beautiful life: Stefan George and his circle in Heidelberg]. In H. Treiber & K. Sauerland (Eds.), *Heidelberg im Schnittpunkt intellektueller Kreise. Zur Topographie der "geistigen Geselligkeit" eines "Weltdorfes"* (pp. 310–327). Opladen, Germany: Westdeutscher Verlag.

Livingstone, D. N. (1995). The spaces of knowledge: Contributions towards a historical geography of science. *Environment and Planning D: Society and Space, 13*, 5–34.

Livingstone, D. N. (2003). *Putting science in its place: Geographies of scientific knowledge.* Chicago: University of Chicago Press.

McClelland, C. E. (1980). *State, society, and university in Germany, 1700–1914*. Cambridge, England: Cambridge University Press.

Meusburger, P. (2008). The nexus between knowledge and space. In P. Meusburger (Series Ed.) & P. Meusburger, M. Welker, & E. Wunder (Vol. Eds.), *Knowledge and space: Vol. 1. Clashes of Knowledge: Orthodoxies and heterodoxies in science and religion* (pp. 35–90). Dordrecht, The Netherlands: Springer.

Meusburger, P. (2009). Milieus of Creativity: The Role of Places, Environments, and Spatial Contexts. In P. Meusburger (Series Ed.) & P. Meusburger, J. Funke, & E. Wunder (Eds.), *Knowledge and Space: Vol. 2. Milieus of Creativity. An Interdisciplinary Approach to Spatiality of Creativity* (pp. 97–153). Dordrecht, The Netherlands: Springer.

Mussgnug, D. (2003). Die "Wiederemporbringung" der Heidelberger Universität 1803 [The renaissance of Heidelberg University in 1803]. In A. Kohnle, F. Engehausen, F. Hepp, & C. L. Fuchs (Eds.), *". . . so geht hervor ein' neue Zeit"*. *Die Kurpfalz im Übergang an Baden 1803* (pp. 131–145). Heidelberg, Germany: Verlag Regionalkultur.

Nauck, E. T. (1956). *Die Privatdozenten der Universität Freiburg i. Br. 1818–1955* [The unsalaried university lecturers of Freiburg University im Breisgau, 1818–1955]. *Beiträge zur Freiburger Wissenschafts- und Universitätsgeschichte* (Vol. 8). Freiburg, Germany: Albert.

Nipperdey, T. (1988). *Religion im Umbruch. Deutschland 1870–1918* [Religion in change: Germany, 1870–1918]. Munich: C. H. Beck.

Ophir, A., & Shapin, S. (1991). The place of knowledge: A methodological survey. *Science in Context, 4*, 3–21.

Pfetsch, F. R. (1974). *Zur Entwicklung der Wissenschaftspolitik in Deutschland 1750–1914* [On the development of science policy in Germany, 1750–1914] (p. 52). Berlin: Duncker & Humblot.

Richarz, M. (1974). *Der Eintritt der Juden in die akademischen Berufe. Jüdische Studenten und Akademiker in Deutschland 1678–1848* [The entry of Jews in academic professions. Jewish students and academics in Germany, 1678–1848]. Tübingen, Germany: Mohr.

Rienhardt, A. (1918). *Das Universitätsstudium der Württemberger seit der Reichsgründung. Gesellschaftswissenschaftliche und statistische Untersuchungen mit einer Darstellung und Beurteilung akademischer Gegenwartsfragen* [University studies in Württemberg since the founding of the Reich: Social science studies and statistical studies with a delineation and evaluation of contemporary academic issues]. Tübingen, Germany: Mohr.

Ringer, F. K. (1979). *Education and society in modern Europe*. Bloomington: Indiana University Press.

Sauerland, K. (1995). Heidelberg als intellektuelles Zentrum [Heidelberg as an intellectual center]. In H. Treiber & K. Sauerland (Eds.), *Heidelberg im Schnittpunkt intellektueller Kreise. Zur Topographie der "geistigen Geselligkeit" eines "Weltdorfes"* (pp. 12–30). Opladen, Germany: Westdeutscher Verlag.

Schmeiser, M. (1994). *Das Berufsschicksal des Professors und das Schicksal der deutschen Universität 1870–1920 ; eine verstehend soziologische Untersuchung* [Career problems of professors and the destiny of the German university 1870–1920: An interpretive study]. Stuttgart, Germany: Klett-Cotta.

Shapin, S. (1998). Placing the view from nowhere: Historical and sociological problems in the location of science. *Transactions of the Institute of British Geographers, 23*, 5–12.

Titze, H. (1987). *Das Hochschulstudium in Preußen und Deutschland 1820–1944 . Datenhandbuch zur deutschen Bildungsgeschichte* [University study in Prussia and Germany, 1820–1944: Data on educational history in Germany] (Vol. 1, part 1). Göttingen, Germany: Vandenhoeck & Ruprecht.

Titze, H. (1995). *Wachstum und Differenzierung der deutschen Universitäten 1830–1945. Datenhandbuch zur deutschen Bildungsgeschichte* [Growth and diversification of German universities 1830–1945. Data on educational history in Germany] (Vol. 1, p. 2). Göttingen, Germany: Vandenhoeck & Ruprecht.

Treiber, H. (1995). Fedor Steppuhn in Heidelberg (1903–1955). Über Freundschafts- und Spätbürgertreffen in einer deutschen Kleinstadt [Fedor Steppuhn in Heidelberg (1903–1955):

Friendship meetings and social gatherings in a small German town]. In H. Treiber & K. Sauerland (Eds.), *Heidelberg im Schnittpunkt intellektueller Kreise. Zur Topographie der "geistigen Geselligkeit" eines "Weltdorfes"* (pp. 70–118). Opladen, Germany: Westdeutscher Verlag.

Turner, S. R. (1987). Universitäten [Universities]. In K.-E. Jeismann & P. Lundgreen (Eds.), *Handbuch der deutschen Bildungsgeschichte: Vol. 3. 1800–1870 . Von der Neuordnung Deutschlands bis zur Gründung des Deutschen Reiches* (pp. 221–249). Munich: C. H. Beck.

Volkov, S. (1987). *Soziale Ursachen des Erfolgs in der Wissenschaft. Juden im Kaiserreich* [Social causes of success in science: Jews in the German empire]. *Historische Zeitschrift, 245,* 315–342.

Walter, S. (2003). *Das Wesen der Religion: Übersetzungsgeschichte und Übersetzungskritik der russischen Feuerbachübersetzung von 1862 aus Heidelberg* [The essence of religion: History and critique of the Russian translation of Feuerbach from Heidelberg in 1862]. Moscow: Mart.

Weber, M. (1984). *Max Weber, ein Lebensbild* (3rd ed.). Tübingen, Germany: Mohr. (Original work published 1926)[Max Weber: A portrait of his life]

Weick, C. (1995). *Räumliche Mobilität und Karriere. Eine individualstatistische Analyse der baden-württembergischen Universitätsprofessoren unter besonderer Berücksichtigung demographischer Strukturen* [Spatial mobility and career: A statistical analysis of the careers of professors in Baden-Württemberg with special reference to their demographic structures]. Heidelberger Geographische Arbeiten 101, Heidelberg: Selbstverlag des Geographischen Instituts.

Weisert, H. (1974). Die Verfassung der Universität Heidelberg. Überblick 1386–1952 [The constitution and structure of Heidelberg University: Overview, 1386–1952]. *Abhandlungen der Heidelberger Akademie der Wissenschaften, Philosophisch-historische Klasse, Abhandlung 2.* Heidelberg, Germany: Carl Winter.

Weisert, H. (1983). *Geschichte der Universität Heidelberg. Kurzer Überblick 1386–1980* [History of Heidelberg University in brief, 1386–1980]. Heidelberg, Germany: Carl Winter.

Wolgast, E. (1986a). *Die Universität Heidelberg 1386–1986* [Heidelberg University, 1386–1986]. Berlin: Springer.

Wolgast, E. (1986b). *Sechshundert Jahre Universität Heidelberg* [600 years of Heidelberg University]. In Rektor der Universität Heidelberg (Ed.), *600 Jahre Ruprecht-Karls-Universität Heidelberg 1386–1986. Geschichte, Forschung, Lehre* (pp. 21–27). Munich, Germany: Länderdienst Verlag.

Wolgast, E. (1987). Phönix aus der Asche? Die Reorganisation der Universität Heidelberg zu Beginn des 19. Jahrhunderts [Phoenix out of the ashes? The reorganization of Heidelberg University at the beginning of nineteenth century]. In F. Strack (Ed.), *Heidelberg im säkularen Umbruch. Traditionsbewusstsein und Kulturpolitik um 1800* (pp. 35–60). Stuttgart, Germany: Klett-Cotta.

Wolgast, E. (1989). Der deutsche Antisemitismus im 20. Jahrhundert [German anti-Semitism in the twentieth century]. *Heidelberger Jahrbücher* (Vol. 33, pp. 13–37). Heidelberg: Springer.

Wolgast, E. (2000). Die badische Hochschulpolitik in der Ära Friedrichs I. (1852/56 bis 1907) [The university policy of Baden in the era of Frederik I (1852/1856–1907)]. *Zeitschrift für die Geschichte des Oberrheins, 148 (New Series 109),* 351–368.

Wolgast, E. (2008). Die Universität Heidelberg zu Beginn des 19. Jahrhunderts [Heidelberg University at the beginning of the nineteenth century]. In W. Moritz (Series Ed.) & F. Engehausen, A. Schlechter, & P. J. Schwindt (Vol. Eds.), *Friedrich Creuzer 1771–1858 . Philologie und Mythologie im Zeitalter der Romantik.* Archiv und Museum der Universität Heidelberg Schriften (Vol. 12, pp. 9–24). Heidelberg, Germany: Verlag Regionalkultur.

Academic Travel from Cambridge University and the Formation of Centers of Knowledge, 1885–1954

Heike Jöns

Academic travel can be regarded as an important process in the production and exchange of scientific knowledge and the formation of scholarly networks across the globe (see, for example, Ackers, 2005; Altbach, 1989; Barnett & Phipps, 2005; Blumenthal, Goodwin, Smith, & Teichler, 1996; OECD, 1996). It may be necessary for accessing field sites, libraries, and archives when producing new scientific arguments, it may contribute to the dissemination and evaluation of scientific knowledge in different places, and it may play an important role for informal contacts, exchanges, and collaborations between distant laboratories and academics. In his seminal book *Science in Action*, Bruno Latour (1987, pp. 210–211, 220–221) pointed out that the circular process of going away, crossing the paths of other people, and returning enables scientists to perform at least three tasks: first, to mobilize new and often unexpected resources for knowledge production; second, to test the value of newly constructed truth claims in different settings; and, third, to spread arguments and facts in time and space. Academic travel from Cambridge University, as studied in this chapter, closely resembles the circular form of mobility and mobilization that Latour (1987, pp. 215–257) considers to be constitutive in the production of scientific knowledge on a variety of scales. In this context, academic travel can be defined as physical journeys by academics for the purpose of research, lecturing, visiting appointments, consulting, and other professional tasks. These journeys may last from a few days to a couple of years, but they are in principle temporary absences, with the traveling academics intending to return to their academic institution.

Geographers and historians of science have devoted considerable attention to scientific travel in the ages of modern discovery and exploration (e.g., Blunt, 1994; Bravo, 1998; Driver, 2001; Duncan & Gregory, 1999; Heffernan, 1994; Hume &

H. Jöns (✉)
Department of Geography, Loughborough University, Loughborough LE11 3TU, UK
e-mail: h.jons@lboro.ac.uk

This chapter is an abridged version of a paper original published in the *Journal of Historical Geography*. It is adapted and republished within the journal's terms and conditions. For the original article, see Jöns, H. (2008). Academic travel from Cambridge University and the formation of centres of knowledge, 1885–1954. *Journal of Historical Geography, 34*, 338–362.

Youngs, 2002; Livingstone, 1992, pp. 102–138; Livingstone & Withers, 1999; Pratt, 1992; Simões, Carneiro, & Diogo, 2003; Sörlin, 1993). However, surprisingly little is known about the nature of academic travel at modern research universities as they emerged in the nineteenth century (for a recent history, see Clark, 2006). Although Livingstone (2003) concluded from historical case studies that "the growth of scientific knowledge has been intimately bound up with geographical movement" (p. 177), the ways in which academic travel contributed to the production of knowledge at modern universities have yet to be explored. In a recent review of the field, Powell (2007) suggests that "an arena in which there is great potential for future contributions by geographers of science is around discussions of travel, instrumentation, and metrology . . . . It is in this work that the competing understandings of the spatiality of science have undertaken fecund interaction" (p. 321; see also Bourguet, Licoppe, & Sibum, 2002). Considering the multiple meanings of travel for academic work, the study of academic travel in fact appears to be a central issue for histories and geographies of knowledge, science, and higher education (for the history of science, see Secord, 2004). Geographical movement of academics, which is at least partly motivated by their work, contributes to the production and dissemination of ideas, arguments, and facts and thus to the alteration of existing knowledge in the places concerned.[1] Accordingly, Barnett and Phipps (2005) maintain that academic travel "indexes more than physical journeys, and physical journeys themselves point to changes in conceptions of knowledge and ideas" (p. 5). Dean (2005) addresses the material effects of academic travel as well, arguing that "processes of circulation like travel produce scientific knowledge and change geographies" (p. 1).

Studying geographical movements of academics therefore helps reveal the wider geographies of academic work and intellectual exchange, and of the knowledge and networks involved, in at least five ways. First, studying academic travel helps explore the important question of where academic knowledge was produced. Crucial in this regard are inquiries about the places that are deemed interesting or resourceful enough to contribute to knowledge production and those that are neglected as sites of study and thus not put onto the agendas of research and teaching. Second, academic travel is inextricably linked to the history of geographical thought, for all travelers, as Heffernan (1994) put it, "were contributing to a synthesizing geographical consciousness in which newly acquired knowledge about the physical and human attributes of particular places and regions gleaned from other disciplines was drawn together under the banner of geography" (p. 22; see also Driver, 2001, pp. 2–3.). Mapping the destinations of academic travel thus also provides an idea about the making of geographical knowledge circulating within and beyond the academy.

Third, by identifying clusters of visitors to certain places, the study of academic travel helps map the hierarchies of centers of knowledge production and dissemination. Which places attracted most academics and for what reasons? For example, destinations of research travel will have promised valuable resources (and might have been exploited for research). Knowledge was transferred, discussed, and exchanged in the context of visiting appointments, invited lectures, and conference travel, whereas the places of academic consulting indicate where scientific expertise was sought by governments, companies, and other bodies. The geographies of academic travel for different types of work thus expose global power relations within science and higher education. Fourth, they also promise further insights into the relationship between science and politics. Mainly a question of data availability, the few studies conducted on geographical and topical patterns of modern academic travel have relied on sponsorship programs. These programs emerged in the course of the nineteenth and twentieth centuries and often contributed to the development of nationally oriented systems of academic patronage (see, for example, Heffernan, 1994; Jöns, 2003; Patiniotis, 2003, pp. 48–49, 51; Teichler, 2002). Analyzing the geographical destinations and topical foci of French international fieldwork funded by the Comite des Travaux Historiques (founded in 1834) and the Service des Missions (founded in 1842) from 1830 to 1914, Heffernan (1994) argued that the traveling scholars' preference for western Europe and the Mediterranean basin was the result of a "compromise between intellectual curiosity, practical expediency and the political judgements of well-briefed scientists and scholars under the direct control of government officials and civil servants" (p. 29). This argument raises questions about the ways in which state authorities or imperial and other political interests have influenced academic travel from modern universities and thus the production and trade of knowledge at institutions of higher education (on the relationship between science and empire, see Bell, Butlin, & Heffernan, 1995; Blunt, 1994; Drayton, 2000; Driver, 2001; Godlewska & Smith, 1994; Harrison, 2005; MacLeod, 1993, 2001).

Fifth, the study of academic travel directs attention to the wider networks that sustain modern universities. Institutional histories of universities rarely track the complex external linkages constitutive of research and teaching. However, these linkages—of which circular movements of academics represent but one dimension—provide important information on the extent and nature of the internationalization of higher education and the wider political, cultural, and intellectual meaning of universities.[2]

In this chapter I begin to explore these larger questions by analyzing travel cultures of Cambridge academics from 1885 to 1954. In the late nineteenth and early twentieth centuries, when Britain was the leading world power, Cambridge University attracted "talent from every corner of the globe" (Brooke, 1993, p. xv). It succeeded in doing so "partly because a prejudice was abroad—not often related to the facts—that Cambridge was a distinguished university" (p. xv). Essentially, the university experienced a "transition to a major international centre of scholarly and scientific teaching and research" (p. xix) from the 1880s to the 1950s, a change that makes it a suitable focus for analyzing the role of academic travel in the formation of the modern university. The analysis is based on all recorded applications for leave of absence by Cambridge University Teaching Officers, whose ranks encompass professors, readers, lecturers, demonstrators, assistant lecturers, and a few other academic posts. This database was created from individual entries in the minute books of the General Board of Studies (since 1926 the General Board of the Faculties), which has been the body responsible for all academic affairs at Cambridge since 1882. Providing information on the date of the application; the applicant's name, position, and subject; and, less frequently, on the period, purpose, and destination of the planned leave of absence, this unique data set permits analysis of both individual itineraries and global patterns of academic travel in different types of work.[3]

This chapter focuses on the global geographies of academic travel by interpreting geographical patterns of physical journeys by Cambridge academics (as far as they can be reconstructed from the archival data). Identifying the places involved in academic travel from Cambridge reveals as much about the making of "geography," or a "geographical consciousness," in Britain and Europe at the time as it tells about the intellectual history of Cambridge University in the late nineteenth and early twentieth centuries. Other epistemological and ontological questions of academic travel, including the metaphorical movement "across the boundaries of fields of knowing" (Barnett & Phipps, 2005, p. 6) and an academic's "personal journey of change, challenge, and even struggle" (p. 6), will inform the analysis as they inevitably intersect with geographical movement. However, rather than looking at the results of specific journeys, this chapter discusses the nature and wider meaning of academic travel for Cambridge University and the global geographies of knowledge, science, and higher education.

By examining transnational linkages of an individual academic institution, this analysis complements national perspectives on the study of science (e.g., Ben-David, 1992; Crawford, 1992; Crawford, Shinn, & Sörlin, 1993). To facilitate meaningful comparisons, however, the maps presented in this chapter display travel destinations by country because this geographical scale was the one most frequently encountered in the sources (as opposed, for example, to places, cities, and supranational regions). The period studied is the 70 years after Cambridge academics were first required to apply for leave of absence during term time. The starting point is therefore defined by the institutionalization of academic travel at Cambridge University and the related availability of comparable travel data; the endpoint coincides with the close of the first decade after World War II, a time that saw the disintegration of the British Empire and the beginning of commercial air travel. The latter development provides particularly good reasons for ending the analysis after a period of 70 years, for the dawn of commercial air transport brought about revolutionary changes for the travel behavior of the individual.[4] The chosen span from 1885 to 1954 also facilitates critical reflection on an argument in the history of cross-boundary science, one that associates that period with the transition from a prevailing nationalism to a stronger internationalism said to have emerged before World War I and to have eventually flourished after World War II (Crawford, 1992, pp. 28–78; Crawford et al., 1993).

Drawing upon recent work in the history and geography of science, this study has four broader aims. First, by mapping the global "reach" of Cambridge University, it strives to reveal the wider geography of academic and geographical knowledge circulating between different places of knowledge production at the time. Second, it explores the role of academic travel for the emergence of Cambridge as a modern research university. Third, it investigates the ways in which the circular flows of academics fostered the formation of knowledge centers elsewhere. Fourth, it examines how the global geographies of academic travel varied among different types of work.

A History of Professionalization

Cambridge University is the younger of England's two ancient universities. Like Oxford, it is characterized by a loose confederation of faculties, colleges, and other bodies, a structure that has led to multiple commitments of its academics to research, teaching, and administration (Brooke, 1993, pp. 20–24, 515–516; Clark, 2006, p. 458; Harrison, 1994). The foundation for a successful research university was laid in the late nineteenth century when a new system of college taxation recommended by the University Commission of 1872, alongside large private endowments, allowed for significant investment in new posts and facilities in the sciences.[5] It included the creation of the renowned Cavendish Laboratory in 1870 and the development of new scientific fields such as electronics and computer technology, nuclear physics, biochemistry, and genetics.[6]

Far beyond Cambridge, this period was associated with the rise of new subjects, professorships, laboratories, journals, and supporting institutions. Through the foundation of scientific associations, the emergence of international conferences, and the implementation of research awards such as the Nobel Prize (granted from 1901 onward), academic networks were promoted well beyond the nation-state, and comprehensive university and college reforms marked important steps in the professionalization of higher education and research (Authier, 1998). In Cambridge, the new University Statutes of 1882 introduced the requirement for professors and readers to "be resident throughout full term time" (University of Cambridge, 1882, p. 61).[7] These new regulations specified the limits of the area within which professors and readers were required to live (defined by distance from the university church) and how often they had to be available to students and colleagues in Cambridge. For any period in which professors and readers were not able to adhere to the strictly defined rules of residence, they had to obtain permission from the General Board. They did so by applying for "leave of absence" that freed an academic, for the period specified in the application or agreed to by the General Board, from the duties as a University Teaching Officer, including residence requirements.[8] Therefore, the data on applications for leave of absence document all absences from university duties during full term time (whether parts or all of it), the majority of which involved travel, and thus physical absence, from Cambridge. As Cambridge academics remained free to travel during vacations, the presented data are far from providing a complete picture of their whereabouts. They do, however, offer unique insights into the destinations of academic travel that would otherwise remain unknown. More important, one can assume that the data cover most journeys exceeding 3 months, for that period is the length of the Long Vacation. Absences from Cambridge longer than 3 months used at least parts of full term time and thus required an application for leave of absence.[9]

In this section, data on all recorded applications for leave of absence are discussed, including absences relating to war service, personal affairs, representation of the university in parliament, or ill health. In the remaining sections, the focus is on applications for academic leave of absence (in short, "academic leave") for research, lecturing, teaching, consulting, administration, and other professional tasks. Particular emphasis is placed on the development of "sabbatical" leave, understood as periodically granted academic leave providing university teachers an opportunity for self-improvement "with full or partial compensation following a designated number of years of consecutive service" (Good, 1959, p. 424; see also Eells, 1962, p. 253; Sima, 2000, pp. 68–69). Freed from all teaching and administrative responsibilities, or rather obliged effectively to give up those duties in Cambridge in order to receive the full stipend, academics on sabbatical leave were to some extent free to decide how to use their time for "professional, personal and creative growth" (Zahorski, 1994, p. 8), but it often involved the concentration on study and research, writing, travel, or visiting appointments. Because the first system of sabbatical leave had been established at Harvard University in 1880, a focus on research and self-improvement, to a lesser extent also on curriculum development and service to the discipline or academic institution, has been considered "an investment in the future of the institution granting [the leave]" (Eells, 1962, p. 253; see also Sima, 2000, p. 73). Sabbatical leave can thus be regarded as an important part of the research culture at a modern university. In Cambridge, this research culture slowly began to emerge in the late nineteenth century through a series of university reforms that, among other things, introduced the first regulations on the residence, accessibility, and academic leave of professors and readers.

In the 7 decades after the institutionalization of academic travel at Cambridge University, the annual number of applications for leave of absence grew considerably, if not steadily (Fig. 1). In the period up to World War I, the General Board, governing the academic affairs of about 100 university professors, readers, and lecturers (Fig. 2), received up to ten applications for leave of absence per year; fewer

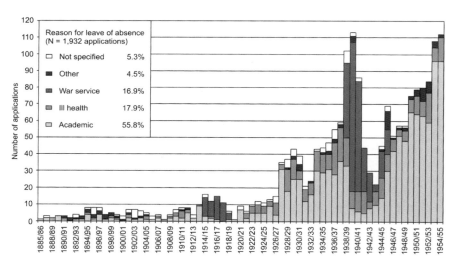

Fig. 1 Applications for leave of absence by Cambridge University Teaching Officers, academic years 1885–1886 to 1954–1955. Adapted from Cambridge University Archives, Minutes of the General Board (GB), GB Min III.1 to GB Min III.7 and GB 160, Boxes 301–307

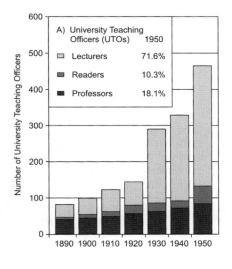
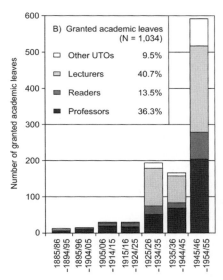

Fig. 2 Cambridge University Teaching Officers and granted academic leaves by decade. Adapted from (**a**) Cambridge University Calendar, 1890, 1899–1900; *Cambridge University Reporter*, 1909–1910, 1919–1920, 1929–1930, 1939–1940, 1949–1950; (**b**) Cambridge University Archives, Minutes of the General Board (GB), GB Min III.1 to GB Min III.7 and GB 160, Boxes 301–307

than half of them for academic reasons. In other words, the idea of spending a sabbatical leave abroad was not well established in Cambridge at the time.

During World War I, a considerable number of Cambridge academics were engaged in war-related services, a fact that underlines the great significance of academic knowledge and expertise in times of war (see, for example, Barnes, 2006; Heffernan, 1996). Over the whole period under study, the General Board conducted a very generous policy on granting leaves of absence. Only 0.9% of all recorded applications for leave of absence were not granted. In 4.6% of all cases, the decision was postponed to get more information from the applicant, to confirm the support of the relevant Faculty Board, or because the applicant had made only a preliminary inquiry to assess the chances of approval. Most applications stated that they had already received consent from the Head of Department or from the Faculty Board so that the decision of the General Board could mostly be regarded as a formal approval of arrangements previously agreed on within the relevant departments and faculties.[10]

Whereas the annual number of applications remained rather low after the war, the jump in applications in the late 1920s relates to new regulations implemented in the University Statutes of 1926. Most important, the new statutes provided rules for the "periodic," or sabbatical, leave as it had become known from American universities (Eells, 1962; Eells & Hollis, 1962.). Every University Teaching Officer was henceforth entitled to take off one term for every 6 terms of service, a formula corresponding to 1 year, or 3 terms, after 6 years of service, with the salary being

agreed on by the General Board (Registry of the University, 1928, pp. 40–41.).
It remained possible to obtain an "occasional" or "additional" leave of absence on
account of illness or any other sufficient cause, such as lecturing in another university or undertaking work on behalf of the government. In these cases the term
in question still counted toward "sabbatical" leave if the duration of absence did
not exceed the greater part of that term. The new rules applied not only to professors and readers as before, but included lecturers, demonstrators, assistant lecturers,
and a few other academic posts whose holders previously had not been obliged to
apply for leave of absence, although some of them had been doing so anyway (see
Registry of the University, 1928, Statute D.XII.1; University of Cambridge, 1955,
pp. 32, 399). In the following decades under consideration, these regulations were
frequently amended and adjusted, particularly in regard to financial arrangements,
but they did not change in principle.

The introduction of sabbatical leave was part of the third major university reform
that helped transform Cambridge into a modern research university. Following
the Royal Commissions on Oxford and Cambridge of 1852 and 1874, a third
Commission, set up in 1919 and chaired by former Prime Minister Herbert
Henry Asquith (1852–1928), initiated new constitutions and revised the statutes
of both universities and their colleges. Building on the Universities of Oxford and
Cambridge Act of 1923 and on the Universities and College Estates Act of 1925, this
body completed the whole process of reform by 1926. The Asquith Commission's
recommendations of 1922 addressed four major areas: university government, the
organization of teaching and research work, the accessibility of the universities and
colleges to poor students, and the place of women in the universities (Brooke, 1993,
pp. 341–369). One of the Commissioners' first concerns was "to ensure that university teachers had sufficient time for research and the instruction of postgraduate
students" (p. 352). However,

> in spite of their strong emphasis on research, the Commissioners of 1922 proceeded rather
> hesitantly in their support of study leave. They refused to propose the institutions 'of a
> "Sabbatical Year"' 'for reasons of public economy', but urged that a fund be set up to help
> meritorious cases for leave. (p. 354)

The fact that the University of Cambridge Statutory Commissioners, who finished what the Royal Commissioners had begun, incorporated both a leave of
absence fund and sabbatical leave in the new University Statutes of 1926 illustrates
the association between research, study leave, and travel.

In other studies of cross-boundary science, the period before World War I has
been associated with the emergence of international academic travel and collaboration. This development was influenced by the "transport revolution" related to
steamship and railway travel and linked to the rise of international standards in methods, units, taxonomies, and equipment (Crawford, 1992, pp. 38–43, 61–64; See also
Crawford et al., 1993, pp. 13–21; Geyer & Paulmann, 2001; Lyons, 1963). Evidence
from Cambridge, however, suggests that only lecturing abroad, primarily in the
United States, played a major role in the academic years 1910–1911 to 1913–1914
and then again from 1920–1921 to 1924–1925. By contrast, the introduction of

the new regulations in 1926 prompted a qualitative difference in the annual number of applications for academic leave, especially for research-oriented sabbaticals. Based on different types of leave of absence, the revised framework provided regular opportunities to concentrate on research and travel for an extended time and thus to plan periods of research in advance. The new rules thereby contributed to the professionalization of research in the ancient university.

The increase of academic travel triggered by the university reform of 1926 and by a growing number of university teachers was interrupted by the impact of the world economic crisis from 1931 to 1933 and ended abruptly at the beginning of World War II. In comparison to the academic year 1937–1938, the number of applications for leave of absence in the academic year 1939–1940 more than doubled. The university released teaching officers at all career stages and in all academic fields in order to engage in military, academic and administrative work related to the war at home and abroad. By May 1940, the number of staff had already decreased so much that the General Board adopted a more restrictive policy on releasing their academic staff for volunteer service and service on demand.[11]

In the war years (1939–1945), travel for academic reasons declined significantly and often had a specific political agenda. In the first postwar decade, however, when the age of travel by ship was slowly being replaced by the age of commercial air transport, academic travel experienced unprecedented growth. In Cambridge, this increase was related to an expanding body of university teachers (particularly at the lecturer level), a growing number of applications for academic leave, and an application process more standardized than it used to be. Academic leave comprised 56.6% of all 1,826 granted applications from 1885–1886 to 1954–1955. The vast majority of these 1,034 academic leaves were granted in the final 3 decades under consideration; the postwar decade alone accounting for over half of them. The relation of granted academic leave per professor and reader rose from 0.2 in the decade from 1885 to 1894 to 2.1 in the postwar decade from 1945 to 1954 (Fig. 2). In the latter period, on average, 23.8% of Cambridge professors were granted academic leave in any one academic year, with an average of 5.7% on sabbatical in each academic year.[12]

The historical development of leave of absence at Cambridge University proceeded from the modest beginnings of institutionalized academic travel before World War I to its boom in the decade after World War II (Fig. 1). The following analysis details how the geographies of academic travel varied over time and among different types of work. It also examines the extent to which the motivations for and destinations of academic travel were influenced by scientific, political, economic, and other interests.

The Geography of Internationalization

Academic leaves by Cambridge University Teaching Officers were divided roughly equally between short-term absences of up to 1 month (33%), medium-term absences of longer than 1 month up to 3 months (35%), and long-term absences of more than 3 months (28%).[13] Whereas long absences increased after World War I, short-term absences slightly increased in the decade after World War II, a change that can be explained by improved transportation and a growing emphasis on short-term lecturing and conference travel. Out of the 1,034 academic leaves granted from 1885 to 1954, 38.8% were sabbatical leaves; 46.0%, additional leaves; and 15.2%, other types of leave. In the decade after World War II, this relation amounted to 40.2% sabbatical leaves, 56.8% additional academic leaves, and 3.0% other academic leaves.[14] At least 81.4% of all granted academic leaves had a destination outside Cambridge, and at least 72.7% involved travel abroad, thus illustrating that the history of academic leave is largely a history of academic travel (Table 1).[15]

Anglo-American Ties

In the late nineteenth and early twentieth centuries, Cambridge academics traveled to a few places scattered around the globe. Archaeologists and scholars of oriental languages conducted research in Italy, Greece, Cyprus, Syria, and Egypt. Conference visits were paid to Germany, Austria, and Italy—all before World War I—and to Australia, Japan, Canada, and the United States. From 1895 onward, a number of Cambridge academics were invited to give lectures at Harvard, Princeton, Yale, Johns Hopkins, and several other American universities. This mobilization of expertise, mainly in the form of invited lectures and visiting appointments, intensified in the following decades and can be regarded as an important contribution to the emergence of American universities as worldwide academic centers.

A key context for this development was the success of the German research university in the second half of the nineteenth century, of which "the rest of Europe and North America had to take note and then action" (Clark, 2006, p. 449.). With the foundation of Johns Hopkins University in 1876, modern research and graduate training in the style of the German university had begun to take shape in the United States, whereas the 1890s "constituted the takeoff decade for the diffusion of the graduate school in America" (p. 453). It seems to be no accident that 7 of the 14 recorded academic travelers from Cambridge in the decade from the academic year 1895–1896 to 1904–1905 went to the United States. On the one hand, the period from 1898 to 1906 has been described as the "formative years of Anglo-American understanding" (Gelber, 1938, p. 1). On the other hand, "many sensed that American graduate programs had attained near or actual parity" by that time (Clark, 2006, p. 463). The related research expertise and newly established laboratories and libraries constituted a major attraction for Cambridge academics, resulting in growing Anglo-American academic ties that flourished on the basis of a common English language.

Bearing these global geographies of higher education and research in mind, one finds that the institutionalization of academic travel at Cambridge University in the 1880s appears to be both the result of pressure for modernization and professionalization, coming partly from the "improved" German university model at the newly

Table 1 Destinations of academic leaves at Cambridge University by decade (in percentage of academic leaves)

Destination	Decade (in academic years)							
	1885–1886 to 1894–1895	1895–1896 to 1904–1905	1905–1906 to 1914–1915	1915–1916 to 1924–1925	1925–1926 to 1934–1935	1935–1936 to 1944–1945	1945–1946 to 1954–1955	1885–1886 to 1954–1955
United Kingdom of Great Britain & Ireland	41.7	7.1	20.7	6.9	11.4	7.3	11.8	11.4
Thereof Cambridge	16.7	0.0	0.0	0.0	4.1	3.0	2.4	2.8
Abroad[a]								
a. British Empire overseas (as of 1914)	50.0	85.7	75.9	89.7	63.2	64.2	77.4	72.7
	16.7	21.4	20.7	24.1	17.6	17.6	14.5	16.2
b. United States of America	16.7	50.0	31.0	48.3	17.6	22.4	31.4	27.9
c. Continental Europe	8.3	21.4	20.7	10.3	21.2	17.6	26.5	23.2
Thereof Germany	0.0	7.1	3.4	0.0	5.2	1.8	4.7	4.2
d. Other places	8.3	0.0	3.4	10.3	8.8	6.7	7.6	7.5
Not specified	8.3	7.1	3.4	3.4	25.4	28.5	10.8	15.9
Number of academic leaves	12	14	29	29	193	165	592	1,034

[a]The sums of the figures in columns 3–9 exceed the given totals because one trip may have included destinations in different categories. Adapted from Cambridge University Archives, Minutes of the General Board (GB), GB Min III.1 to GB Min III.7 and GB 160, Boxes 301–307.

established U.S. research universities, and a contribution to the development of an Anglo-American academic hegemony in the twentieth century.[16] This situation also means that the idea of the modern research university as first embodied by the German university model was imported to Cambridge via the United States. The relative paucity of the links between the competing knowledge centers of Germany and England is underlined by the small number of Cambridge academics who traveled to the world-wide centers of academic knowledge in Germany before 1945 and particularly in the decade after World War I (Table 1).[17]

From 1925 to 1944, academic travel from Cambridge University not only flowed increasingly to the United States but also intensified within Europe and extended to British colonies in the west Atlantic, in Africa, and in south and southeast Asia (Fig. 3). In the first 2 decades of the twentieth century especially, Cambridge academics often followed the routes of imperial power, much as their French counterparts supported by the Service de Mission did, thus underlining the close relationship between scholarship and the European imperial project in that period.[18] The post-1945 decade, however, saw the emergence of an increasingly postcolonial transnational exchange (Table 1). Despite the global reach of travel destinations, the increasing internationalization from the pre-World War II to the post-World War II decade was characterized by a disproportionate rise in academic travel to the United States, which demonstrated the existence of a closely linked Anglo-American world by the mid-twentieth century. In the decade after 1945, decolonization and European integration eventually made academic travel to continental Europe much more important than that to the British Empire.[19]

The following analysis of travel for different types of academic work reveals in more detail the ways in which asymmetrical power-relations between different places were articulated in the global flows of Cambridge academics.

Asymmetrical Power Relations

Cambridge academics often traveled for a mix of reasons, combining lecture tours with conference visits, research with consulting, or academic tasks with private affairs. Despite the complex nature of motivations for academic travel, it is possible to identify five main types of work that show distinct geographical patterns of destinations at the global level: research and travel (35.4%), visiting appointments (9.9%), lecturing (17.9%), conference visits and representation (18.2%), and administration and consulting (10.4%).[20]

Travel for the purpose of research and learning aimed at a variety of places across the globe (Fig. 4a). By assembling heterogeneous resources such as samples, artifacts, collaborators, knowledge, and ideas in different parts of the world in order to build new knowledge claims, Cambridge academics contributed to making their university a Latourian center of calculation. According to Latour (1987), the recurring mobilization of "anything that can be made to move and shipped back home" (p. 225) in scientific centers of calculation such as the university, the laboratory,

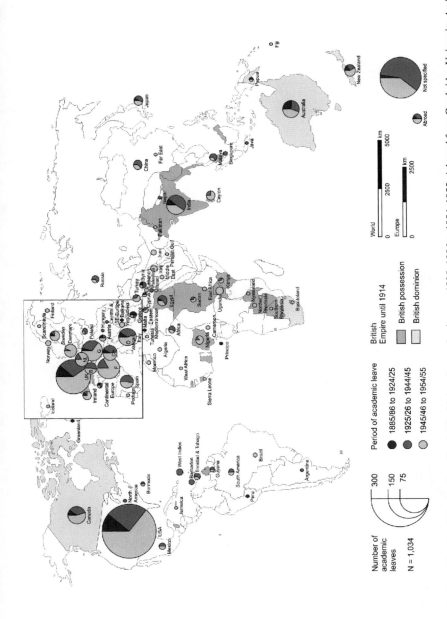

Fig. 3 Destinations of academic leaves from Cambridge University, academic years 1885–1886 to 1954–1955. Adapted from Cambridge University Archives, Minutes of the General Board (GB), GB Min III.1 to GB Min III.7 and GB 160, Boxes 301–307

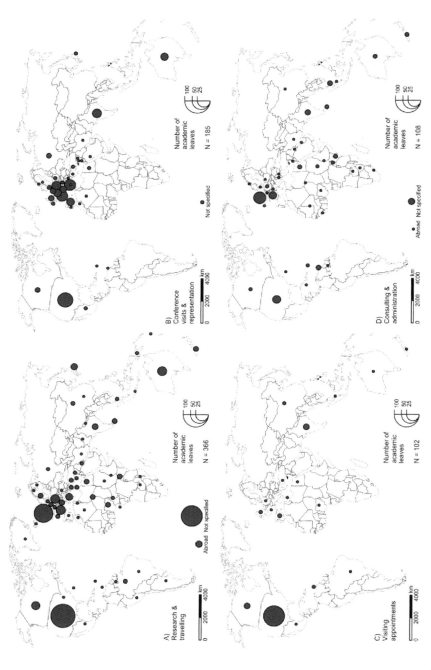

Fig. 4 Destinations of academic leaves from Cambridge University by different types of work, academic years 1885–1886 to 1954–1955. Adapted from Cambridge University Archives, Minutes of the General Board (GB), GB Min III.1 to GB Min III.7 and GB 160, Boxes 301–307

the archive, and the museum shaped the cumulative character of European science from the ages of discovery and exploration and established Europe as the center of the imperial age. Essentially, each full "cycle of capitalization" (p. 218) created an advantage in knowledge that made distant places familiar and thus controllable.

From this perspective, the University of Cambridge can be conceptualized as a site of knowledge production and dissemination that was constituted and maintained to a considerable extent by the travels of its academics.[21] Representing a central institutional node within the global networks of academia, Cambridge University was embedded within a complex set of overlapping processes of mobilization by different centers of calculation. Exemplified by increasing travels of Cambridge academics to the United States, these travels not only served the travelers' experiences and resources but also helped enhance the prestige and expertise of the new American research universities in the eyes of the world. More than two thirds of all visiting appointments held by Cambridge scholars in the academic years from 1885–1886 to 1954–1955 were at American universities (Table 2). American universities were able to offer these temporary posts of 3–12 months because they possessed the necessary financial resources to attract foreign scholars, and the prospect of learning about the latest research infrastructure was inviting even for established Cambridge academics. At the time, these circumstances were to be found in only a few other places of the world, most of which were located within the British Empire. Only five visiting appointments were held at the competing centers of knowledge production across continental Europe—none of them in the natural and technical sciences (Fig. 4c).

The regional clusters formed by global geographies of lecturing resembled those of conference travel (Fig. 4b). Characterized by short-term absences, more than half of the conference travel by Cambridge academics was to established European research centers. For this type of transnational academic exchange, destinations in the United States were less important than those in the British Empire and other places, including India, Australia, and Russia. Closely linked academic and political interests were particularly evident with several lecture tours and conference visits to the Middle East in the postwar period, often arranged through the British Council. In July 1946 the permanent undersecretary of the Foreign Office, Orme Sargent, identified the Middle East as a region of prime importance for the future of the British Commonwealth, an assessment that resulted in a concerted policy aimed at boosting Britain's image in the region, in competition with not only the Soviet Union but also the United States and France (Vaughan, 2005, p. 152).

Cambridge academics were also involved in a number of consulting and administrative jobs for the British government, for colonial governments, and for international organizations and corporations (Fig. 4d). In the postwar decade, for example, Cambridge academics advised on Jewish Education in Palestine (1946), nutritional problems in the British Zone of Germany (1946), and land drainage in Poland (1947). They gave advice on the sugar industry in British Guiana (1949), helped the New Zealand government institute the new constitution of Western Samoa (1949), and commented on proposals for a higher college for Africans in the British Central African Territories (1952). In 1954, Cambridge academics helped the Anglo-Iranian

Table 2 Destinations of academic leaves at Cambridge University by type of work (in percentage of academic leaves)

Destination	Type of work						
	Research and travel	Visiting appointment	Lecturing	Conference and representation	Administration and consulting	Unspecified sabbatical leave	Total 1885–1886 to 1954–1955
United Kingdom of Great Britain & Ireland							
	14.5	1.0	16.0	3.8	23.1	2.4	11.4
Thereof Cambridge	6.3	0.0	0.0	0.0	3.7	2.4	2.8
Abroad[a]							
a. British Empire overseas (as of 1914)	66.1	99.0	83.5	95.1	70.4	0.0	72.7
	15.6	19.6	10.6	15.7	38.0	0.0	16.2
b. United States of America	26.2	71.6	38.3	20.0	10.2	0.0	27.9
c. Continental Europe	15.6	4.9	33.5	52.4	16.7	0.0	23.2
Thereof Germany	3.6	1.0	7.4	7.0	1.9	0.0	4.2
d. Other places	12.0	4.9	3.2	8.1	7.4	0.0	7.5
Not specified							
	19.4	0.0	0.5	1.1	6.5	97.6	15.9
Number of academic leaves							
	366	102	188	185	108	85	1,034

[a]The sums of the figures in columns 2–8 exceed the given totals because one trip may have included destinations in different categories. Adapted from Cambridge University Archives, Minutes of the General Board (GB), GB Min III.1 to GB Min III.7 and GB 160, Boxes 301–307.

Oil Company with legal problems arising from the reopening of the Iranian oil fields, served as economic advisors to the Iraq Development Board, and investigated an outbreak of disease in poultry in Egypt. In 1955, they inquired into a threatened labor dispute in the oil industry of Trinidad, made recommendations for land reform in Ceylon, and advised the government of Sierra Leone on the reconstruction of their cost-of-living index.

If the expertise created by research and teaching at Cambridge had greatly benefited from colonial ties, these examples illustrate how it was used, in turn, in the running of the British state and empire, the reconstruction of postwar Europe, and the creation of favorable postcolonial relations within the British Commonwealth. Academic and political interests often went hand in hand when Cambridge academics traveled abroad. Their journeys were an important chapter in Europe's appropriation of non-European regions that contributed to strong colonial and, later, postcolonial ties.[22]

Conclusion

Academic travel from Cambridge University from 1885 to 1954 contributed significantly to the formation of modern research universities as global centers of knowledge production and dissemination. Cambridge benefited from the mobilization processes of its academics in various ways, including the production of new knowledge claims; the access to and import of new ideas, prestige, objects, and infrastructure; the links to academic networks and prospering research centers; and the raising of funds for the university. Other institutions, such as the new research universities in the United States, began to attract visiting Cambridge academics from the 1890s onward and emerged as prominent centers of science and learning in the first half of the twentieth century. The institutionalization of academic travel at Cambridge through a major university reform in the 1880s can be seen as both the result of a need for modernization in the face of the newly established U.S. research universities and as a contribution to the development of an Anglo-American academic hegemony in the twentieth century.

Although a part of the modern research culture in the United States since the 1880s, the idea of the "sabbatical" leave did not arrive in Cambridge until 1926, during another major university reform that explicitly promoted research work. The related professionalization and internationalization of academic travel was shaped by intensifying Anglo-American relations and changing power geometries of world politics. Destinations in the British Empire figured prominently in the first 3 decades of the twentieth century, but their importance decreased with decolonization, whereas European reconstruction after 1945 was accompanied by growing academic ties with the continent. The strong interrelation of academic expertise and political interests was also underlined by academic travel to regions of strategic importance to Britain, such as Africa and the Middle East, and by consulting work for colonial institutions and postcolonial supranational organizations.

Comparing the geographies of academic travel for different types of work has revealed distinct geographical patterns of interaction on a global scale. Based on the empirical analysis presented here, my argument is that the global flows of Cambridge academics were shaped in at least four ways. First, they were inextricably linked to international politics. Second, they were produced by a complex set of overlapping mobilization processes by different centers of calculation. Third, they were molded by the geographically uneven distribution of cultural, symbolic, social, and economic capital. Fourth, they were influenced by varying spatial relations of the practices conducted at the travel destinations. In order to improve the understanding of the geographies of knowledge, science, and higher education, it is particularly important to examine further the relationship between academic travel and the social and material specificities of different scientific practices.

In conclusion, this study has shown that the rise of a research culture at modern universities went hand in hand with an emphasis on academic travel. Academic travel can thus be regarded as a constitutive element in the formation of modern research universities as centers of knowledge production. As Lux and Cook (1998) point out, "viewing science in action situated in one location . . ., without taking account of the many people coming and going to and from other sites, overlooks something essential" (p. 211). The existence of an Anglo-American academic hegemony since the mid-twentieth century illustrates that these essential global geographies of academic travel, particularly with regard to socially important practices such as visiting appointments, lecturing, and conference travel, also indicate the rise and intensification of academic networks and thus help anticipate future changes in the geographies of knowledge, science, and higher education.[23]

Acknowledgments I wish to thank the Alexander von Humboldt Foundation, the University of Nottingham, and the Cambridge University Archives for supporting this work. I am particularly grateful to Mike Heffernan for hosting my stay as a Feodor Lynen Postdoctoral Research Fellow at the School of Geography, University of Nottingham, and for providing invaluable input and encouragement at all stages of this project. Alexa Färber, Tim Freytag, Michael Hoyler, Phil Hubbard, and three anonymous referees also kindly provided thoughtful comments on earlier versions of this chapter, and Felix Driver's careful editorial comments greatly improved the final text. This abridged version of the article originally published in the *Journal of Historical Geography* also benefited from David Antal's suggestions as the technical editor of the book series. Documents from the Cambridge University Archives are reproduced with kind permission of the Syndics of Cambridge University Library.

Notes

1. This interest is closely linked to recent work on historical and geographical variations in the production and dissemination of scientific knowledge. See Harris (1998), Livingstone (1995, 2000, 2002, 2003), Naylor (2005), Ophir and Shapin (1991), Shapin (1995), Smith and Agar (1998), Withers (2001).
2. The standard history of the University of Cambridge in the twentieth century (Brooke, 1993) contains a chapter entitled "The University and the World, 1945–1990: A cosmopolitan society" (pp. 511–566). Though the text sketches some transnational influences without referring to travels by Cambridge academics, it does mention "the extensive travels of many British

students [that] help to explain the rapid waves of fashion from distant places which have flowed into Britain" (p. 520).

3. Data was supplemented from various sources, including personnel files available in the Cambridge University Archives, the "List of Cambridge University Officers" accessible via Research tools on the Janus Homepage (Cambridge University Library Catalogues), the Oxford Dictionary of National Biography (Online Edition, 2004, Cambridge University Library), and the *Who Was Who* (Online Edition and printed volumes, Cambridge University Library). In addition, letters of correspondence on leaves of absence occasionally survived in the personnel files and in the minute books of the General Board.

4. In the early 1950s, Cambridge academics rarely flew but went to the United States and other places by sea. In fact, it was not until 1958 that the number of passengers traveling by air over the Atlantic Ocean first exceeded the number traveling by sea. The same year saw the world's first scheduled transatlantic flights, which resulted in fundamentally changing patterns of geographical mobility (Beaty, 1976, pp. 241–243; see also Rimmer & Davenport, 1998, p. 125). I am grateful to Lucy C. S. Budd for pointing me to this literature.

5. As of 1882, Cambridge colleges were required "to give funds for university lectureships and chairs" (Clark, 2006, p. 459; see also Darwin, 1993, p. 357).

6. According to Clark (2006), "the modern era began, academically and belatedly, with the appointment of the Graham Commission as royal visitor to Oxford and Cambridge in 1850" (p. 457), but the resistance to the work of the commission was particularly strong at Cambridge. Even in 1872–1873, the Master of Sidney Sussex "might still accuse the reformers of drawing their inspiration from German universities" (p. 458), arguing that "the university's only function was to conduct examinations, and that all teaching, including the support of laboratories, could be managed by the colleges" (p. 458).

7. Cambridge University's academic year is divided into 3 terms of slightly varying length: Michaelmas (October–December), Lent (January–March), and Easter (April–June). The Long Vacation usually lasts from July to September. Residence was obligatory in about three quarters of each term, that is, in roughly 45 out of 60 days (the Easter term being a bit shorter than the others). See Statute B.XI.2 in University of Cambridge (1882, p. 61).

8. From the academic year 1885–1886 onward, professors and readers had to send residence returns to the General Board every term, notifying that body of (a) absences from home for more than two nights or (b) the weeks in which he or she had not spent the required hours in Cambridge. Leave of absence was given to any professor or reader, "provided that the General Board of Studies are of opinion that such leave may be granted without prejudice to the interests of the University" (*Cambridge University Reporter*, 1885, p. 931); Draft, for the General Board of Studies, 22 February 1886, Cambridge University Archives, General Board of Studies, Min III.1, p. 109.

9. This ruling applied to journeys of those academics who had to apply for leave of absence during full term time. Before 1926 these persons were professors and readers; as of 1926 this group encompassed all Cambridge University Teaching Officers.

10. In special circumstances that exceeded the regulations and thus the competence of the General Board, the Council of the Senate was asked to sanction a "Grace." This procedure was particularly frequent during World War I. Between September 1939 and October 1947, emergency statutes empowered the General Board to grant leave of absence for as long as the applicant was engaged in national service.

11. On April 23, 1940, 18 out of 42 University Teaching Officers between 25 and 30 years of age were engaged in national service. Another 11 University Teaching Officers in that age bracket were liable for military service, and 13 academics were reserved in other occupations. Minutes, April 24, 1940, Cambridge University Archives, General Board of Studies 160, Box 302, p. 184.

12. According to Sima (2000, p. 70), the share of professors on sabbatical in a given year reportedly hovers around 5% at public universities in the United States and may have reached 20% at private elite institutions (1990s).

13. The remaining 4% of applications did not state a period of absence or left the period open. The differentiation of short-term, medium-term, and long-term absences aims to provide the bigger picture. It seems to be impossible to find coherent categories for a period of 70 years that was characterized by rapid changes in transport and communication. Moreover, the period of absence may have been longer than stated in the application when the leave of absence preceded or followed vacations.

14. In the context of the professionalization of the application process after 1926, cancellations of granted academic leaves were recorded in the minutes of the General Board because such nullification involved changes in the applicants' rights to sabbatical leaves. This information suggests that about 5% of the granted academic leaves were cancelled (whole period: 4.6%; decade after World War II: 4.9%).

15. These figures refer to the minimum because the destination of 15.9% of all academic leaves was not specified. The data show that at least 2.7% of all academic leaves were spent only in Cambridge.

16. As Clark (2006) notes, "In the 1880s, contemporaries [in Britain] sensed a revolution was taking place. Every university of importance, it was said, must have facilities for science" (p. 459).

17. In an analysis of the career paths of European academics, Taylor, Hoyler, and Evans (2008) find no major linkages between German and English universities during the nineteenth century.

18. According to Heffernan's (1994) work on the Service de Mission between 1870 and 1914, about "a third of all post-1870 missions involved either a colonial or an Eastern Mediterranean destination" (p. 34; see also pp. 28–29 in that source and the corresponding figures in Table 1 of this chapter).

19. In the academic years from 1885–1886 to 1954–1955, the ten most important countries for academic travel from Cambridge University were the United States (27.9%), Great Britain (10.7%; Ireland: 1.0%), France (4.5%), Germany (4.2%), India (3.4%), Canada (3.0%), Italy (3.0%), Australia (2.6%), Switzerland (2.0%), and the Netherlands (1.7%).

20. The first two categories were dominated by sabbatical leaves; the last three categories mainly comprised additional leaves. Unspecified sabbatical leaves account for the remaining 8.2% of academic leaves.

21. Whereas the idea of scientific centers of calculation has been widely debated, developed, and criticized by geographers, this chapter offers an opportunity to explore empirically the types and geographies of those circulatory flows constitutive of a particular scientific center of calculation. For related geographical studies see Barnes (2006), Bravo (1999), Gregory (2000, pp. 316–317); and Heffernan (2000).

22. Bell et al. (1995) stress that both modern European science and European imperialism "were supremely ambitious, universalizing projects concerned to know all, to understand all and, by implication, to control all" (p. 3).

23. For example, increasing academic travel from and to China may contribute to the emergence of a new apex of knowledge production centers in the twenty-first century (Jöns, 2007; Leydesdorff & Zhou, 2005).

References

Ackers, L. (2005). Moving people and knowledge: Scientific mobility in the European Union. *International Migration, 43*, 99–131.

Altbach, P. G. (1989). The new internationalism: Foreign students and scholars. *Studies in Higher Education, 14*, 125–136.

Authier, M. (1998). Zeittafel [Timeline]. In M. Serres (Ed.), *Elemente einer Geschichte der Wissenschaften* (pp. 946–1030). Frankfurt am Main: Suhrkamp.

Barnes, T. J. (2006). Geographical intelligence: American geographers and research and analysis in the Office of Strategic Services, 1941–1945. *Journal of Historical Geography, 32*, 149–168.

Barnett, R., & Phipps, A. (2005). Academic travel: Modes and directions. *The Review of Education, Pedagogy, and Cultural Studies, 27*, 3–16.

Beaty, D. (1976). *The water jump: The story of transatlantic flight.* London: Secker & Warburg.

Bell, M., Butlin, R., & Heffernan, M. (1995). Introduction. In M. Bell, R. Butlin, & M. Heffernan (Eds.), *Geography and imperialism, 1820–1940* (pp. 1–12). Manchester, England: Manchester University Press.

Ben-David, J. (1992). *Centres of Learning: Britain, France, Germany, United States.* New Brunswick, England: Transactions Publishers.

Blumenthal, P., Goodwin, C. D., Smith, A., & Teichler, U. (Eds.). (1996). *Academic mobility in a changing world: Regional and global trends.* London: Jessica Kingsley Publishers.

Blunt, A. (1994). *Travel, gender, and imperialism: Mary Kingsley and West Africa.* New York: Guilford Press.

Bourguet, M.-N., Licoppe, C., & Sibum, H. O. (Eds.). (2002). *Instruments, travel and science: Itineraries of precision from the seventeenth to the twentieth century.* London: Routledge.

Bravo, M. T. (1998). Precision and curiosity in scientific travel: James Rennell and the orientalist geography of the new imperial age, 1760–1830. In J. Elsner & J. P. Rubiés (Eds.), *Voyages and visions: Towards a cultural history of travel* (pp. 162–183). London: Reaktion Books.

Bravo, M. T. (1999). Ethnographic navigation and the geographical gift. In D. N. Livingstone & C. W. J. Withers (Eds.), *Geography and enlightenment* (pp. 199–235). Chicago: University of Chicago Press.

Brooke, C. N. L. (1993). *A History of the University of Cambridge: Vol. 4: 1870–1990.* Cambridge, England: Cambridge University Press.

Cambridge University Reporter. (1885, June 23). Grace No. 4, p. 931.

Clark, W. (2006). *Academic charisma and the origins of the research university.* Chicago: University of Chicago Press.

Crawford, E. (1992). *Nationalism and internationalism in science, 1880 – 1939 : Four Studies of the Nobel population.* Cambridge, England: Cambridge University Press.

Crawford, E., Shinn, T., & Sörlin, S. (1993). The nationalization and denationalization of the sciences: An introductory essay. In E. Crawford, T. Shinn, & S. Sörlin (Eds.), *Denationalizing science: The contexts of international scientific practice* (pp. 1–42). Dordrecht: Springer.

Darwin, J. (1993). The growth of an international university. In J. Prest (Ed.), *The illustrated history of Oxford University* (pp. 336–370). Oxford, England: Oxford University Press.

Dean, K. J. (2005). *Settler Physics in Australia and Cambridge, 1850–1950.* Unpublished Ph.D. dissertation, University of Cambridge, England.

Drayton, R. (2000). *Nature's government: Science, imperial Britain, and the 'improvement' of the world.* New Haven, CN: Yale University Press.

Driver, F. (2001). *Geography militant: Cultures of exploration and empire.* Oxford, England: Blackwell.

Duncan, J., & Gregory, D. (Eds.). (1999). *Writes of passage: Reading travel writing.* London: Routledge.

Eells, W. C. (1962). The origin and early history of sabbatical leave. *Bulletin of the American Association of University Professors, 48*, 253–256.

Eells, W. C., & Hollis, E. V. (1962). *Sabbatical leave in American higher education: Origin, early history and current practices.* Washington, DC: Government Printing Office.

Gelber, L. M. (1938). *The rise of Anglo-American friendship: A study in world politics, 1898–1906.* London: Oxford University Press.

Geyer, M. H., & Paulmann, J. (Eds.) (2001). *The mechanics of internationalism.* Oxford, England: Oxford University Press.

Godlewska, A., & Smith, N. (Eds.). (1994). *Geography and empire.* Oxford, England: Blackwell.

Good, C. (Ed.). (1959). *Dictionary of education*. New York: McGraw-Hill.

Gregory, D. (2000). Cultures of travel and spatial formations of knowledge. *Erdkunde, 54*, 297–319.

Harris, S. J. (1998). Long-distance corporations, big sciences, and the geography of knowledge. *Configurations, 6*, 269–304.

Harrison, B. (Ed.). (1994). *The history of the University of Oxford: Vol. 8. The twentieth century*. Oxford, England: Oxford University Press.

Harrison, M. (2005). Science and the British Empire. *Isis, 96*, 56–63.

Heffernan, M. (1994). A state scholarship: The political geography of French international science during the nineteenth century. *Transactions of the Institute of British Geographers, New Series, 19*, 21–45.

Heffernan, M. (1996). Geography, cartography and military intelligence: The Royal Geographical Society and the First World War. *Transactions of the Institute of British Geographers, New Series, 21*, 504–533.

Heffernan, M. (2000). Mars and Minerva: Centres of geographical calculation in an age of total war. *Erdkunde, 54* 320–333.

Hume, P., & Youngs, T. (Eds.). (2002). *The Cambridge companion to travel writing*. Cambridge, England: Cambridge University Press.

Jöns, H. (2003). Geographies of international scientific exchange in their political context: The case of visiting scholars to Germany in the second half of the twentieth century. In J. Nemes Nagy & A. Jakobi (Eds.), *Frontiers of geography* (pp. 227–247). Budapest: Department of Regional Geography, Eötvös Loránd University.

Jöns, H. (2007). Transnational mobility and the spaces of knowledge production: A comparison of global patterns, motivations and collaborations in different academic fields. *Social Geography, 2*, 97–114.

Latour, B. (1987). *Science in action: How to follow scientists and engineers through society*. Cambridge, MA: Harvard University Press.

Leydesdorff, L., & Zhou, P. (2005). Are the contributions of China and Korea upsetting the world system of science?. *Scientometrics, 63*, 617–630.

Livingstone, D. N. (1992). *The geographical tradition*. Oxford, England: Blackwell.

Livingstone, D. N. (1995). The spaces of knowledge: Contributions towards a historical geography of science. *Environment and Planning D: Society and Space, 13*, 5–34.

Livingstone, D. N. (2000). Making space for science. *Erdkunde, 54*, 285–296.

Livingstone, D. N. (2002). *Science, space and hermeneutics* (Hettner Lectures, Vol. 5). Heidelberg: Department of Geography.

Livingstone, D. N. (2003). *Putting science in its place: Geographies of scientific knowledge*. Chicago: University of Chicago Press.

Livingstone, D. N., & Withers, C. W. J. (Eds.) (1999). *Geography and enlightenment*. Chicago: University of Chicago Press.

Lux, D. S., & Cook, H. J. (1998). Closed circles or open networks? Communicating at a distance during the scientific revolution. *History of Science, 36*, 179–211.

Lyons, F. S. L. (1963). *Internationalism in Europe, 1815–1914*. Leiden, The Netherlands: A. W. Sythoff.

MacLeod, R. (1993). Passages in imperial science: From empire to commonwealth. *Journal of World History, 4*, 117–150.

MacLeod, R. M. (Ed.). (2001). *Nature and empire: Science and the colonial enterprise*. Chicago: University of Chicago Press.

Naylor, S. (2005). Historical geographies of science: Places, contexts, cartographies: Special issue on the historical geographies of science. *British Journal for the History of Science, 38*, 1–12.

OECD (Ed.). (1996). *Internationalisation of higher education*. Paris: OECD.

Ophir, A., & Shapin, S. (1991). The place of knowledge: A methodological survey. *Science in Context, 4*, 3–21.

Patiniotis, M. (2003). Scientific travels of the Greek scholars in the eighteenth century. In A. Simões, A. Carneiro, & M. P. Diogo (Eds.), *Travels of learning: A geography of science in Europe* (pp. 47–75). Dordrecht: Springer.

Powell, R. C. (2007). Geographies of science: Histories, localities, practices, futures. *Progress in Human Geography, 31*, 309–329.

Pratt, M. L. (1992). *Imperial eyes: Travel writing and transculturation.* London: Routledge.

Registrary of the University (Ed.). (1928). *Statutes of the University of Cambridge and Passages from Acts of Parliament Relating to the University.* Cambridge, England: Cambridge University Press.

Rimmer, P. J., & Davenport, S. M. (1998). The geographer as itinerant: Peter Scott in flight, 1952–1996. *Australian Geographical Studies, 36*, 123–142.

Secord, J. A. (2004). Knowledge in transit. *Isis, 95*, 654–672.

Shapin, S. (1995). Here and everywhere: Sociology of scientific knowledge. *Annual Review of Sociology, 21*, 289–321.

Sima, C. M. (2000). The role and benefits of the sabbatical leave in faculty development and satisfaction. *New Directions for Institutional Research, 105*, 67–75.

Simões, A., Carneiro, A., & Diogo, M. P. (2003). *Travels of learning: A geography of science in Europe.* Dordrecht: Kluwer Academic Publishers.

Smith, C., & Agar, J. (Eds.). (1998). *Making space for science: Territorial themes in the shaping of knowledge.* Basingstoke: Macmillan.

Sörlin, S. (1993). National and international aspects of cross-boundary science: Scientific travel in the 18th century. In E. Crawford, T. Shinn, & S. Sörlin (Eds.), *Denationalizing science: The contexts of international scientific practice* (pp. 43–72). Dordrecht: Springer.

Taylor, P. J., Hoyler, M., & Evans, D. M. (2008). A geohistorical study of "the rise of modern science": Career paths of leading scientists in scientific networks. *Minerva, 46*, 391–410.

Teichler, U. (Ed.). (2002). *ERASMUS in the SOCRATES programme: Findings of an evaluation study.* Bonn: Lemmens.

University of Cambridge (Ed.). (1882). *Statutes of the University of Cambridge with some acts of Parliament relating to the university.* Cambridge, England: Cambridge University Press.

University of Cambridge (Ed.). (1955). *Ordinances of the University of Cambridge: To 1 October 1955.* Cambridge, England: Cambridge University Press.

Vaughan, J. (2005). 'A certain idea of Britain': British cultural diplomacy and the Middle East, 1945–1957. *Contemporary British History, 19*, 151–168.

Withers, C. W. J. (2001). *Geography, science and national identity: Scotland since 1520.* Cambridge, England: Cambridge University Press.

Zahorski, K. J. (1994). *The sabbatical mentor: A practical guide to successful sabbaticals.* Bolton, MA: Anker Publishing.

Part III
Designing Knowledge Spaces

Big Sciences, Open Networks, and Global Collecting in Early Museums

Dominik Collet

The traditional narrative of early modern science focuses on a small circle of grand inventors. It stresses the revolutionary crux of the discoveries they made in the solitude of their laboratories. Consequently, this older narrative puts priority on scientific fields that can be pursued in relative autonomy, such as astronomy, physics, optics, and mathematics.

Recent research, however, has started to question the limitation to these "small sciences" and the narrow groups of gifted scholars who worked on them. Attention has broadened to include the agents, correspondents, assistants, and patrons involved in scientific practice. Along with this shift, the "big sciences," such as natural history, geography, and cartography, have begun to attract heightened scholarly interest. These disciplines involve large numbers of people working collaboratively in different locales across great distances (Harris, 1998). The new awareness of the social dimension of early modern science has also led to a reappraisal of the effect of space and spatiality on scientific developments, with wide networks and global connectedness constituting a key field of investigation. In this chapter I examine early modern museums as nodes of this global scientific encounter and discuss the potential and the problems of assembling a microcosm of the material world and of utilizing it for research on distant parts of the planet.

One of the challenges of the big sciences has been to communicate and construct scientific knowledge at long range, with reports on exotic flora and fauna or foreign landmarks having to cross large geographical and, sometimes, cultural distances. In the past, many people involved in these networks did not know each other personally. Methods to establish the validity of knowledge claims by people of untested credibility were therefore crucial. For that reason two aspects dominate the social history of early modern science. One is the spatial distribution, or geography, of science and the communication flow between its different locales. The other is the position and construction of "witnesses."

D. Collet (✉)
Seminar für Mittlere und Neuere Geschichte, Georg-August-Universität Göttingen, 37073 Göttingen, Germany
e-mail: dominik.collet@phil.uni-goettingen.de

P. Meusburger et al. (eds.), *Geographies of Science*, Knowledge and Space 3,
DOI 10.1007/978-90-481-8611-2_6, © Springer Science+Business Media B.V. 2010

Lux and Cook (1998) have proposed a model that integrates both facets. Borrowing from Granovetter (1973), they contrast two extremes: (a) closed circles based on the "strong ties" of personal acquaintance and (b) open networks that rely on the "weak ties" of correspondence and long-distance communication. Although it is usually assumed that closed circles of initiates work more efficiently than open networks do, Lux and Cook observe that weak, but flexible, networks were often the more efficient alternative for disseminating and, crucially, verifying scientific knowledge claims. They tried to show that the exclusive, almost secretive circle of the Académie des Sciences in Paris, for example, could not match the wide network of correspondents the Royal Society of London had at its disposal. Lux and Cook even propose that it was not the brilliant individual inventor but rather the increase in communication flows through open networks that was pivotal in the scientific expansion in sixteenth and seventeenth century Europe.

In this chapter I test Lux and Cook's (1998) hypothesis on another field: the early modern museum. These museums constituted spaces in which long-distance communication, natural history, and distributed collaborative work converged. Much of what they exhibited originated in far-away countries. Many of their natural history specimens were at the heart of the biological, zoological, and geological debates of their time. Indeed, several collections were created explicitly to facilitate scientific discourse. In these institutions the gathering of material evidence was intended for the critical review of written reports from distant places. Consequently, communication with overseas witnesses, agents, and donors figured strongly in the daily life of European collectors.

Early modern museums are thus excellent subjects for a study on the modes of a distributed science. They created a condensed geographical space where objects and people from a wide background met, where information was processed, and where new forms of knowledge were disseminated to visitors and guests. Such an enterprise required myriad people for gathering, shipping, categorizing, and arranging these objects. If open networks did have an edge over the closed circles of specialists, it should be observable in the museum environment.

Just as Lux and Cook (1998) choose London's Royal Society for the Improving of Natural Knowledge to illustrate their model, I take the Royal Society's *Musaeum*, or repository, as the basis of the following passages. I also follow their choice to concentrate on the second half of the seventeenth century, the same period during which the Society's museum was in active use.

A Research Museum

The first European "museums" developed in the second half of the sixteenth century. Unlike earlier collections, these *Kunstkammern*, repositories, or cabinets arranged their exhibits according to elaborate taxonomic systems. They were also more accessible and were planned not only with entertainment but also with education in mind. However, they differed substantially from the modern public museum. Very few of the collections were institutionalized; most of them belonged to private "virtuosi." The visitors usually came from a narrow social class, and early museums were also far less specialized. Most collections ranged from objects of art, such as painting, sculpture, and antiques, to monstrous births, miraculous animals, exotic plants, and ethnographic material. In fact, the rising influx of non-European material in the sixteenth century was one of the main reasons for the establishment of these collections (Collet, 2007).

Most museum owners were content to display objects that the popular travel reports praised as strange, rare, or beautiful. They aimed to illustrate the established pool of knowledge and provide an inviting space for social intercourse and friendship. Some of the collectors, however, pursued more ambitious plans. They envisioned their museums not only as places of collecting and sociability, but as centers of research. This group saw the study of the expanding material world and the hitherto unknown products of America and Asia as a source of "new" knowledge that would form the basis of an equally "new" science. The most vocal advocates of this approach were the fellows of the Royal Society of London.

When the Royal Society for the Improving of Natural Knowledge was founded in 1660, the establishment of a museum, or repository, was high on their agenda (Hunter, 1989, pp. 123–155). It suited their view of science as based on indisputable "facts" rather than philosophical speculation. They hoped that the natural objects would help eliminate the divisive discussions that plagued the scientific debates of their day and English post civil-war society in general. Because the Royal Society's membership included Royalists and Puritans, Cartesians and Platonists, the fellows discouraged debate and favored a scientific practice geared to fact-finding rather than interpretation (Schaffer & Shapin, 1985).

The English Civil Wars (1642–1651) had also disrupted earlier collecting activities. Although museums were well established in many parts of the continent by the mid-seventeenth century, only a few such institutions could be found in England. Furthermore, a museum in the trust of a scientific society was without precedent even in Italy or France. The fellows therefore had to invent their own guidelines by which to run their collection. They envisioned the museum primarily as a place where the material world could be examined. The curator Robert Hooke (as cited in Waller, 1705) proudly proclaimed that the "Collection of all varieties of Natural Bodies" would enable them to "read the Book of Nature" rather than the reports of the ancients (p. 338). He saw the use of such a research collection not in "Divertisment, and Wonder, and Gazing, as 'tis for the most part thought and esteemed" (p. 338). Instead, he hoped for an instrument to put the classical books on natural history and the more recent accounts of explorers to the test. The other fellows shared his notion that the rapid influx of objects from overseas had outdated the classical natural histories. They were also concerned about the incongruent and often dubious information in the contemporary travel reports. A new natural history, one based on the museum's exotica and including the whole world now known to them, was therefore agreed on as a primary goal of the Society.

For this task they could initially fall back on a collection they had bought in 1665 from a London entrepreneur (Hunter, 1989, pp. 129–136). Though this museum accentuated exotic natural objects, the exhibits proved to be a haphazard sample of popular monstrous and bizarre curiosities or objects associated with people of renown. They mirrored the tastes of the famous European collections rather than the needs of what the fellows had in mind as the natural history of the world. As a result the Royal Society had to resort to other methods to gather the needed material.

Global Collecting

Europe's virtuosi could usually draw on a large and well-established trade in curiosities to fill their collections (Findlen & Smith, 2002). Although the direct exchange between gentlemen was highly treasured, the vast majority of objects reached the museums through specialized dealers and agents. The Royal Society, however, tried to avoid London's curiosity shops. One reason was the high price of exotic collectables and the Society's desire to keep the membership fees from deterring potential newcomers. Another was the uncertain credibility of simple tradesmen and people of low social status.

Instead, the fellows embarked on an ambitious project to establish their own network of trusted witnesses. Virtuosi from all corners of the world were requested to furnish objects and information. At first, this task was to be administered by a special "correspondence committee." However, most of the work fell on the shoulders of the Society's secretary, Henry Oldenburg. He consulted the relevant travel reports and excerpted long lists of questions and desired objects relating to various regions of the earth. He translated them, had them printed in the *Philosophical Transactions of the Royal Society* or as broadsheets, oversaw their distribution, and collected the returning answers and—occasionally—specimens. Close to a hundred of these lists have survived.[1]

The questionnaires, or "inquiries," went to respected contacts in the Bermudas, the Bahamas, Virginia, and other English colonies. Others were addressed to the Governor of Bombay, the president of the English East India Company in Surat, the "English Consul at Aleppo," and the "English Agent at Ispahan" in Persia.[2] But Oldenburg did not limit his queries to English dominions. Via contacts in Lisbon he hoped to reach Portuguese America. Other correspondents were asked to forward queries to Japan, Lapland, Russia, Ethiopia, and "ye Spanish and French dominions" on the American continent.[3]

Some of the addressees were personal acquaintances of Oldenburg; some were relatives of other fellows. But most of the correspondents were linked to Oldenburg through "weak ties". He knew these correspondents only because of their administrative positions, or from hearsay. Occasionally, Oldenburg seems to have sent out letters to people neither personally nor institutionally known to him, relying on the odd chance that the messages would find their way and elicit a reply. In the case of Brazil, a country the fellows were intensely interested in, the slightest mention of a possible contact could prompt a letter from Oldenburg. When a Lisbon correspondent hinted at a mysterious Brazilian Jesuit scholar, the fellows quickly drew up long lists of sought-after curiosities, without so much as knowing the intended recipient's name. They even elected two other Portuguese academics as overseas members on the same account.[4]

After just a few years, Oldenburg's willingness to communicate with correspondents outside the fellows' personal circles had created a large, "open" network. It had spread from London to most of the colonial world, covering the English overseas settlements and trading posts and a large part of the colonial possessions of other European powers. Specialized and detailed questionnaires had been drawn up for dozens of areas. The various lists were printed, translated, and distributed to a host of senior colonists, diplomats, amateur scientists, and merchants of repute. Given the limited resources of the Royal Society in its early years, the establishment of this comprehensive network was certainly an impressive achievement.

However, a closer look at the actual return of answers and objects shows a rather different picture. Only a small number of reports and specimens ever reached the fellows. Moreover, the information they contained was disparate, poor in quality, and scientifically almost irrelevant. Their influence on the work of the Royal Society remained marginal.

Several factors contributed to this failure. Long-distance communication in the seventeenth and early eighteenth centuries was beset by structural difficulties. Letters got lost and parcels disappeared. Piracy and the frequent quarrels between colonial rivals also posed a serious problem. The few specimens that did reach the fellows were often spoiled and broken. Live plants or animals and objects that were either heavy or large could not be transported in any case. Furthermore, the transmission of exhibits and information took at least several months. The long communication cycles and the rapidly changing colonial world complicated repeated inquiries and limited the exchange on many biological topics or astronomical events.

The unreliability of the network's structure, however, was exceeded by the unreliability of the colonial correspondents. Many of them simply repeated established stereotypes on the "otherness" of extra-European flora and fauna and its native population. Several "eye-witnesses" confirmed that the "Indians" regularly reached the biblical age of 120 years and agreed that some were as tall as giants. Richard Stafford on the Bahamas backed up the many stories about gruesome exotic poisons by an account of poison ivy being able to peel the skin off those unfortunates who looked at it from a distance.[5] Thomas Harpur (as cited in Hall & Hall, 1965–1986, Vol. 3) answered the fellows' questions on the "present studies of ye Persians, and what kind of learning they now excel in" (p. 340) with a stereotypical reference to the Oriental's indulgence in "money-making and voluptuousness" that prevented the locals from "examin[ing] natural things" (p. 467). The Jesuit Jerónimo Lobo confirmed the sighting of the unicorn in Ethiopia and the missing feet of Birds of Paradise (Birch, 1756–1957, Vol. 2 [1756], p. 314). Colonists from North America corroborated the fellows' suspicions that the Africans' dark skin was due to their "dark blood."[6]

The trouble with misleading information was aggravated by its constant repetition. At the outset, the London fellows had all agreed that communicating with correspondents of untested credibility required special care. Because the weak ties that connected them to the fellows could not guarantee their trustworthiness, it was deemed "altogether necessary, to have confirmations of the truth of these things from several hands, before they can be relyed on" ("Inquiries," 1666, p. 415). However, the "facts" confirmed by several correspondents were often based on stereotypes, whereas the few original observations usually came from a single source only. In many cases the "witnesses" in question had simply read the same volumes of travel literature (Collet, 2007, pp. 113–132).

The transmission of specimens did little to mitigate the shortcomings of written communication through weak ties. Most objects conformed to a narrow standard of established collectables. Very few correspondents seemed to bother with the meticulous and laborious instructions of the inquiries at all. They sent what they knew to be popular in European museums. Rhinoceros horns, birds of paradise, and Indian "idols" kept flooding into the repository. Instead of broadening the collection, the donations duplicated the material already on hand. The focus remained firmly on "strange" material with a high visual appeal, often associated with aphrodisiacal or poisonous effects or the capacity to show off the alien nature of the indigenous population. At the same time, the information accompanying the individual objects was scant or nonexistent. Names and places of origin where usually missing, as were all particulars on their local use or the people who had collected or created them. Many donors obviously failed to understand the concept of standardized inquiries or a research collection and its use for empirical science. They were confident that the necessary information could easily be retrieved from printed books, the erudition of which they judged to be far superior to their own.

The uncertainties of long-distance communication; the unreliable, ignorant, and untrained correspondents; the influence of stereotypes; and the narrow canon of established exotic curiosities were not the only problems the fellows had to face. A substantial number of correspondents actively withheld information from the Londoners. They sent established rarities to divert the Royal Society's attention away from information that could compromise their colonial interests. In 1664 Philippo Vernatti, a fiscal of the Dutch East India Company, sent a crate of material in reply to one of the Royal Society's questionnaires. The consignment included a miraculous stone said to cure snake bites, a bird's nest used by "lecherous Chinamen" as an aphrodisiac, and an "odde piece of wood, naturally smelling like human Excrement, used by the Natives ag[ain]st evill Spirits and Incantations."[7] Vernatti's gifts clearly mirrored the taste of European collectors for the bizarre and unusual with an emphasis on the otherness of the non-Christian indigenous population. As an expert in overseas trade, he knew the European virtuosi's fare very well.

The fellows' original queries had indeed asked for curiosities. But they had also asked for maps of the area, for information on the flow of the tides, medicine, and promising raw materials. Accordingly, the Londoners had prepared a reply, thanking Vernatti for his splendid donation but asking him for more information on virtually everything he had sent them. Again they added questions concerning cartography and geography. Vernatti replied curtly that, though he was happy to send them curiosities, he found himself unable to divulge any trade secrets crucial to the Dutch colonial enterprise.[8] The majority of the fellows, however, had already taken his bait. Even before Vernatti's reply, they published a list of his marvelous rarities in *Philosophical Transactions of the Royal Society* and sent copies to correspondents worldwide to encourage similar gifts. Although they had received nothing that went beyond their own established preconceptions of the "foreign" world, they soon forgot about the ominous lacunae in Vernatti's communication.

Because Vernatti was in the employ of a competing colonial power, his behavior was perhaps unsurprising. Far less expected were the similar tactics used by some of the Royal Society's own overseas members. In 1663 the fellows had admitted John Winthrop, Jr., the governor of Connecticut, into their ranks while he was visiting London. Upon his redeparture for the New World, Oldenburg passed him a list of Vernatti's East-Indian gifts and asked him to provide the corresponding "chief raritys" of the West Indies.[9] After several lost, stolen, or miscarried parcels, a package from Winthrop finally arrived in 1669. As in Vernatti's case, it contained typical *Kunstkammer* material: a bird's nest formed in the manner of male genitalia, rattlesnake skins, the "head of a deare, which seemeth not an ordinary head."[10] When the Londoners asked him for details on locations or the indigenous names, Winthrop readily admitted his own ignorance. He claimed that he had received the specimens from "Indians" of the "remote Inland partes little discovered"[11] and was unable to press them for facts.

As with Vernatti, the fellows initially tried to coax information on cartography and mineral resources from Winthrop. But the governor was only too aware that the "Royal" Society was very close to the monarch and his colonial administration. Oldenburg himself had proudly informed Winthrop that the king had taken great pleasure in viewing the strange bird's nests that had arrived from the New World.[12] Furthermore, by the time the fellows pressed Winthrop for maps and minerals, the Duke of Norfolk's fleet had taken the neighboring Dutch colony of New Amsterdam. Royal commissioners had arrived on the ships in order to inspect Connecticut's taxable resources, such as mines and smelting works. Moreover, the founding of the new colony New York in 1664 had added to Connecticut's numerous border conflicts and had made the drawing of maps a delicate enterprise. Winthrop was acutely aware that a natural history specifying resources and geography could well have dire political consequences in the colonial arena (Collet, 2007, p. 292).

The reservations of correspondents on fringes added another challenge to the fellows' communication network. Although the exchange between the Londoners and John Winthrop was one of the few that went through several cycles and could draw on a mutual understanding of scientific codes, it yielded little information. Winthrop carefully stuck to harmless curiosities that he knew would appeal to the sizable group of fellows who were ambivalent about the official "experimentalist" agenda. As in Vernatti's case, his rarities acted as decoys that successfully managed to steer the Londoners away from more delicate topics.

Making Sense

The drawbacks of the Royal Society's open network based on weak ties had serious repercussions for their scientific work. In the beginning, the fellows had hoped to bypass the problems of practicing "science at a distance" by transporting the foreign world into their own. Hooke and others anticipated that the objects in the repository would moderate the known imperfections of written communication. In fact, they expected the objects to replace the study of learned books with a science based on facts and experiment.

However, they soon learned that the spectacular accomplishments of the small sciences could not be repeated in other fields. The bare objects neither delivered conclusive "proof" nor managed to replace the testimony of a trustworthy colleague. And though they had planned their natural history to be practical and useful, the selection of curiosities they received concentrated on the unusual and bizarre. Most of them were shown at a meeting and then permanently shelved in the repository. Very few of them were ever considered for any further use.

Performing "experiments," however, proved to be almost impossible. The many supposedly gruesome poisons they received had little or no effect on various cats and dogs. Because all of the animals escaped the tests unharmed and the various substances on trial were poorly documented, the fellows achieved no conclusive results. Was it not known that the Indians kept their murderous recipes secret from all Christians? Was the potion the fellows had obtained genuine? Faced with the loopholes in the documentation of the objects, they regularly decided that it was the object that was "wrong," not the garish stories (Carey, 2003; Collet, 2007, pp. 302–305). Although the substances in question appeared harmless, the fellows turned out to be incapable of refuting the existence of the many powerful poisons that figured so strongly in the mutually corroborative reports of travelers and explorers.

Missing documentation also caused problems with plant and animal specimens. As taxidermy was still in its infancy, most organic objects that reached the Society were already in various stages of decomposition or had been reduced to their relatively durable parts. When in 1699 the fellow John Ray finally received a batch of dried American plants after years of negotiation, he informed the Society: "But alas! I find as I told you, that I can make but poor work with them."[13] He complained that "neither the colour or figure of the flower" nor the fruit or seeds and "nothing of the root" could be discerned. In his letter he added bitter remarks that "those that gathered them might easily have given an account of all these, as also the place where they were found" and concluded that the plants had left him "in a wilderness and a great uncertainty."

The fellows at first hoped that this "wilderness" could be addressed if written testimonials and the objects flowing through their network were to be combined. They hoped that the questionnaires drawn up for that purpose would afford a framework to standardize information, facilitating comparisons and cross-checks. Few of the questionnaires were ever returned, though, and even they contained only sketchy information rarely based on autopsy or the objects in question. In short, they failed to mitigate the problem. Instead, the questionnaires confirmed established narratives, inadvertently following the very sources the fellows had used during their compilation.

Faced with the decontextualized objects that emerged from this open network, the fellows resorted to consulting travel reports. In order to fill the gaps in the documentation, they copied from the very books they had wanted to examine in the first place. The *Historia Plantarum*, which John Ray published for the Society, was largely a compilation of other books. It did contain some original observations on European flora, but its entries on overseas plants rested primarily on second-hand information. The same was true for much of the museum catalogue published in 1686. Because the fellows had received most of the exhibits without any additional information, the catalogue's author, Nehemiah Grew, had to excerpt liberally from established travelogues. The antidotal qualities of the "unicorn" bird, the sympathetic power of crocodile fat to heal crocodile bites, and the Indians' natural affinity to poisons and insidious murder were all included unchecked. Many of these particulars had figured in the inquiries as doubtful and in need of confirmation. In the end all of them made it into the Royal Society's publications for lack of alternative information, even though the relevant questions had never been answered.[14]

In the early eighteenth century the grand design of a natural history of the world based on the evidence of the museum specimens was unceremoniously abandoned. The problems of this "big science", based as it was on the collaboration of strangers across vast distances, were all too obvious when compared to the manageable and decisive results of physics or mathematics. The Royal Society's museum quickly developed into a salon rather than a laboratory. It served as a meeting place for fellows and foreign virtuosi and became a major tourist attraction (Hatton, 1707, Vol. 2, pp. 666–668; Colsoni, 1699, p. 12). As the reports of its visitors show, the donations had firmly imprinted the established canon of curious collectables on the ambitious research collection. No one noticed any differences between the repository and the traditional collections of "Divertisment, and Wonder, and Gazing." Instead of a research museum, which would diffuse the fruits of experimental science, the repository had grown into a cabinet of curiosities. It illustrated rather than tested the established knowledge. In the following years it quickly lost its place in a society increasingly set on "experimentalism." After long years of neglect, its remains were finally sold to the British Museum in 1779 (Hunter, 1989, p. 154).

Museums, Big Sciences, Open Networks

The fellows' experience was far from unique. Many other collectors who tried to open a direct path to the curiosities of the new world struggled with the same predicament. In Rome the famous Athanasius Kircher possessed excellent contacts to Jesuit missionaries all over the world, but his museum benefited little from his acquaintances.[15] In the German states several collectors also adopted ambitious plans to obtain exotica. The Duke of Saxe-Gotha sent an ambassador to the fabled Christian kingdom of Ethiopia, and the ruler of Hesse-Hanau even acquired a colony of his own to plunder its rarities. In London groups of virtuosi even sponsored expeditions to the Caribbean and North America. None of these projects produced significant results (Collet, 2007, pp. 238–245, 316–318; Smith, 1994, pp. 141–172).

Instead most of the European collections were filled with exotica acquired through the booming and well-established commercial networks, even those of princes and kings. The trade in non-European curiosities had quickly reached even the remotest parts of the continent. Wealthy collectors often received unsolicited catalogues advertising popular exotica from the ports of Amsterdam or Lisbon (Collet, 2007, pp. 79–80). The dominance of the trade network did much to narrow and canonize the group of acceptable collectables from overseas, with most merchants acquiring only those objects that were sure to sell in Europe. It also limited the scientific potential of these objects, for trade was a one-way affair. Asking questions up the line of retailers, seamen, tradesmen, and agents was almost impossible.

Scientifically minded collectors therefore frowned at this piecemeal acquisition through trade. They had intended a museum that would form the hub of a giant wheel, with spokes extending to learned men in all parts of the earth. Along these spokes, information and objects would flow, all converging at the museum and forming a microcosm of the known world. This world at home would allow the observer to study the globe without leaving the comforts of his study.

A look at the practical side of global collecting, however, reveals a less tidy picture. The image of spokes and wheel suggests continuous relations, which were in truth a rare exception. Even if a contact could briefly be established, the backchannel almost never worked. Moreover, the spokes were far from straight. The acquisition of an object usually required many people instead of just two. Each person involved acted as a gatekeeper who selected suitable objects and decided which information was expendable and which needed to be passed on. By the time the objects finally reached the European museum, they lost most of their original contextual documentation.

The communication flows were complicated even further by the collector's attempts to recover at least some of the lost information. Whereas the object would travel along established commercial channels, the contextual facts came from explorers and colonists and reached the collectors after a detour via Europe's printing presses. Objects and information moved along separate routes. Quite often, this geographical separation extended into a chronological disconnection as well. Collectors usually preferred older, established travelogues to new and untested reports. Many of the popular textbooks remained in use for generations, a practice that contributed to the unchanging presentation of the non-European world in European museums at a time when the colonial world itself was changing rapidly.[16]

Harris (1998) has proposed to draw imaginary thread maps by attaching strings to each object of a museum to retrace its way into the collection. He expects these strings to form straight "lines converging to a central node rather like a spider's web" (p. 273). From the collector's perspective, however, such a map would look highly irregular, with broken filaments, some threads leading nowhere, some unraveling into separate strands, and others existing only ephemerally.

The example of the Royal Society also raises questions about Lux and Cook's (1998) findings on the success of open networks. On the European scale, the flexible and fast communication between weak ties was indeed able to surpass the limited exchanges between close friends. Another contact would quickly fill any gaps in the system. The participating men of letters were also able to rely on a common understanding of scientific codes and widely shared verification methods. Outside Europe, open networks worked far less efficiently. A single break in communication was liable to end the reception of news from a whole continent. The low overall number of contacts hampered and often precluded efforts to reroute communication to other channels. It also limited the chances of securing confirmation from a second source.

These structural shortcomings, however, were not the only challenge for intercontinental communication. Different conceptions of science and its methods gave rise to further obstacles. Few of the overseas correspondents knew how to deal with the new instrument of the standardized questionnaire (Harris, 1998, pp. 283–284). Fewer still shared the notion that the authority of objects and experiment surpassed that of the printed word.

These different concepts often stemmed from social distance between the London academics and the colonists, but geographical distance was another important factor. Colonial correspondents like John Winthrop and Philippo Vernatti pursued different goals than the fellows in the metropolis. Life in the colonial sphere encouraged new loyalties, which could result in strategic noncooperation and carefully filtered communication.[17] These correspondents understood that the exchange was never planned as a mutual enterprise. The term *network* suggests equality between its participants. However, what the fellows envisioned was a strictly hierarchical system in which information would flow in one direction only. They lamented the lack of a stable back channel, not because they wanted to share results but because they wanted to ask for even more information to be sent their way.

Even more obvious is the exclusion of the indigenous population from the open network. Though its members were expected to provide much of the material, they were not meant to receive anything in return. Much of their intelligence was in fact used against them. It was twisted in order to reinforce stereotypes of otherness, to associate them with deviant practices, and to legitimate their oppression. Granted, the structural problems of the collector's networks effectively prevented a large-scale appropriation of indigenous knowledge, but the plans nevertheless highlight the inherent inequalities of the open system.[18]

Geographical distance was clearly more than a mere impediment to an otherwise universal scientific truth. It shaped and influenced the participant's fundamental concept of scientific observation (Livingstone, 2003, p. 142). The spatial and cultural divide between different continents and geographical terrains was arduous to cross and even more arduous to recross for the purposes of verification. Innovative open communication networks did not necessarily alleviate the uneven distribution of knowledge across space; they could just as well deepen the divide.

Even though many European scholars were aware of the problems, they did not address them effectively. Nor did they extend their networks to people of low social status. "Indians," seamen, and even most tradesmen, despite their sound practical knowledge, remained outside the collectors' circles. Furthermore, the learned gentlemen did not leave the comfort zones of their homes, museums, and laboratories in order to study nature in situ. Instead, some of them concentrated on the small sciences that operated according to unchanging universal laws. Even in physics or optics the conclusive assertion of truth claims had proved demanding (Shapin, 1994). When it came to challenging natural history, such attempts met with even greater adversity. Whereas physics could be researched in the laboratory regardless of place, natural history differed from country to country. The carefully controlled environment of the workshop, so crucial for experimental science, was patently missing in the collecting of exotic specimens. An experiment could be repeated, transported, and witnessed by large numbers of observers, but natural history remained more or less stationary. Its observation outside its original habitat was often inconclusive and lacked generally accepted methods of verification and demonstration. The fellows' ambivalence about travel reports and about the authority of written knowledge illustrates this uncertainty.[19] It is not surprising that historians of science who have focused on the small sciences have fashioned a clear revolutionary narrative, whereas studies of the messy business of natural history and the complex, intricate geography of science underscore slow change and continuity.

Conclusion

During the seventeenth century many European collectors attempted to establish their museums as global nodes of knowledge. They set up far-reaching networks to gather information and objects from all corners of the known world. Whereas the majority of these collectors were content to use their museums to illustrate established knowledge, some hoped to set up a new type of research collection.

The plans to harvest museum specimens for empirical studies, however, soon faced substantial difficulties. The well-established network of commerce constituted the most readily available source for objects. But though the merchants could supply a steady stream of exotic rarities, their network was ill suited to the task of generating the documentation needed. The piecemeal process of acquisition, shipping, and retail encumbered the transmission of contextual information. Because the number of collectors interested in such documentation was limited, the merchants had little incentive to change. The collectors, in turn, were reluctant to extend their networks past the respected merchant elite to the agents of colonial trade. Many believed that such contacts lacked the necessary credibility and risked blurring the division between the fragile position of the virtuosi and the laboring classes.

Instead, the collectors attempted to create their own open network. The Royal Society devoted much of its resources to this project and managed to establish contacts with a number of like-minded individuals across the globe. However, although such an open network succeeded at intensifying scientific exchange in Europe, it turned out to be inappropriate for overseas communication. The benefits of weak ties—quick diffusion through multiple and flexible channels—were quickly offset by the problems of communication outside the European arena. The correspondence was slow and often erratic, and its few regional participants were hard to reach for extended exchanges of information.

Improved shipping and postal systems did reduce some of the structural problems of the open networks in the eighteenth and nineteenth centuries. However, not all the flaws were due to the specific hurdles confronting early modern transnational communication; some of them plagued all such systems, regardless of scope and time. One such fault was the unreliability of correspondents in a hierarchical system. No "open" network proved truly egalitarian. The uneven distribution of power tended to provoke fringe correspondents into filtering their input correspondingly. Some of them would simply refuse contact, ignore instructions, or even send bogus information. Others chose a strategy of "loyal noncooperation" and sent carefully chosen decoys to protect their own interests.

Second, the detection of such casual or tactical disinformation was not necessarily facilitated by an open system. Lux and Cook (1998) assume that the sheer amount of parallel communication in open networks would quickly expose any incongruities. The fellows, however, learned that it seldom did. In fact, the observations that reached them from multiple sources were usually deceptive. The reports most often repeated relied on hearsay or literature rather than autopsy.

Both defects—the consistent confirmation of stereotypes and the twisted communication between uneven partners—are exacerbated in a colonial setting. They are, however, part of any open network. It seems unlikely that open networks in Europe routinely led to improved methods of verification. Without fundamental changes in the concepts of proof, experiment, and witnessing, increased communication along weak ties could well have resulted in an increased dissemination of established scholastic knowledge.

Instead, the early museums hint at the laborious and fragile business of circulating references (Livingstone, 2003). Latour (1999, pp. 24–79; 2005, pp. 121–140), in his network theory, has therefore expanded the focus from the structure of networks to the way its contents are organized. He has drawn attention to the importance of the unbroken chains of reference required by scientific enquiry that is conducted across large distances. Data traveling along such chains need to remain traceable in both directions in order to successfully support truth claims. The networks of collectors certainly produced a notable "downstream" flow of objects, but they failed to deliver retraceable links between exhibits and their places of origin. Objects and their documentary information often traveled along separate routes, leaving to chance the matter of their eventual reunion. The "upstream" activity (asking questions up the lines of trade or exploration) was even more precarious.[20] Without functioning referential chains, the "actants" (objects and collectors alike) remained unable to transform matter into new knowledge claims, regardless of whether their network was open, closed, strong, or weak.

Although the utilization of museum networks for the purposes of research did not yield the anticipated results, the collections were quite successful on other accounts.

As Burke (2000, pp. 12–17) has stressed, there is no single set of ever-expanding knowledge in the early modern world. Alongside academic knowledge, which is often given precedence by modern researchers, there exist pools of political, technical, or mercantile knowledge. The museums may have failed as a laboratory surrogate or a storehouse of scientific references. But they have succeeded at visualizing, popularizing, and disseminating traditional learning, overseas commerce, and the political dominance of the old world.

This effect is hardly startling, for all museums are inherently conservative enterprises, much better at preserving and presenting than at researching (Sommer, 1999, p. 370). However, through the use of objects instead of the written word, Europe's early modern museums broadened access to traditional forms of knowledge far beyond the narrow ranks of academics. Their rooms served as a crucial meeting place where a limited group of scholars mingled with a socially exclusive, but surprisingly manifold, group of wealthy gentlemen and curious artisans (Collet, 2007, pp. 331–332). They did so irrespective of the traditional barriers of gender, nationality, and religion. In this fashion they contributed to the broad interest in the material world in later decades.

This examination of the global networks of early modern collectors thus reveals certain inadequacies of long-distance trade systems for empirical investigation. It illustrates the indispensable, but fragile, links between object-based research and careful written documentation through unbroken chains of references. Finally, it highlights the vulnerability and the subtle inequalities of "open" networks. Such obstacles are apt to be neglected in an attempt to establish a coherent alternative to the narrative of scientific revolution. Instead, they could serve to sharpen scholarly attention to the predicaments of constructing science at a distance.

Notes

1. Royal Society Archives, London, Classified Papers XIX.
2. Letters of Henry Oldenburg, writing in London, to Sir Robert Boyle (March 13, 1666), Richard Norwood (October 24, 1666), Sir Richard Oxenden (April 6, 1667), Charles Stafford (November 16, 1668), and Charles Hotham (March 7, 1670), in Hall and Hall (1965–1986, Vol. 3 [1966], pp. 57–60, 276–278, 384–385; Vol. 5 [1968], pp. 174–175; and Vol. 6 [1969], pp. 535–536).
3. Hall and Hall (1965–1986, Vol. 5, pp. 54–59, "ye Spanish and French dominions": p. 175; Vol. 8 [1971], pp. 33–34, 155; Vol. 10 [1975], pp. 558–559) and Royal Society Archives, London, Classified Papers XIX, no. 72.
4. Letters of Robert Southwell to Henry Oldenburg (Lisbon, March 6, 1669), and of Henry Oldenburg to António Álvarez da Cunha (London, April 13, 1668) and Gaspar Mere de Souza (London, December 8, 1669), in Hall and Hall (1965–1986, Vol. 4 pp. 313–316; Vol. 5, pp. 433–435; and Vol. 6, pp. 358–359). The Jesuit scholar has been identified as Valentin [E-]Stancel. He wrote several letters to Athansius Kircher, the founder of a famous museum in Jesuit College of Rome, but seems to have sent no objects to his museum.
5. Royal Society Archives, Classified Papers XIX, no. 78, fol. 160; and letter of Richard Stafford to Henry Oldenburg (Bermuda, July 16, 1668), in Hall and Hall (1965–1986, Vol. 4 [1967], pp. 550–553).

6. Letter of Martin Lister to Henry Oldenburg (York, June 27, 1675), in Hall and Hall (1965–1986; Vol. 11 [1977], pp. 373–374).
7. Royal Society Archives, Classified Papers XIX, no. 9.
8. Royal Society Archives, Classified Papers XIX, no. 2, 3, and 9.
9. Letter of Henry Oldenburg to John Winthrop (London, May 26, 1664), in Hall and Hall (1965–1986, Vol. 2, p. 149).
10. Letter of John Winthrop to Henry Oldenburg (Boston, October 4, 1669), in Hall and Hall (1965–1986, Vol. 6, pp. 253–257).
11. Letter of John Winthrop to Henry Oldenburg (Boston, August 26, 1670), in Hall and Hall (1965–1986, Vol. 6, p. 143).
12. Letter of Henry Oldenburg to John Winthrop (London, April 11, 1671), in Hall and Hall (1965–1986, Vol. 7, p. 568).
13. These words and the quotations in the rest of the paragraph are from a letter of John Ray to Hans Sloane (Black Notley, March 23, 1699), in British Library, Ms. Sloane 4037, fol. 246. The Royal Society often sent out detailed, but fruitless, instructions on the gathering, preserving, and documentation of collectables. See, for example, Hall and Hall (1965–1986, Vol. 4, p. 168).
14. The respective inquiries were prompted by passages in Piso (1658, pp. 91, 282, 325). The fellows' inquiries (translated and edited) concerning these subjects can be found in Hall and Hall (1965–1986, Vol. 8 [1971], pp. 147, 242). Piso's remarks were finally copied unaltered into Grew (1686, pp. 42, 65).
15. Close links between Kircher's museum exhibits and his correspondents have often been claimed. However, even though thousands of Kircher's letters and several museum catalogues exist, these assertions remain largely hypothetical. See, for example, Harris (1998, p. 273) and Moser (2001). Kircher's correspondence is available at http://kircher.stanford.edu (retrieved May 11, 2009).
16. A case in point is Piso (1658), which inspired many of the inquiries by the fellows of the Royal Society. Even though it was full of misrepresentations and distorting stories of alterity, or "otherness," it remained in use well into the eighteenth century. See Teixeira (2004).
17. Harris, for the sake of his argument, has stressed the effective control of the nation-state's colonial administration, but others have emphasized the colonial personnel's multiple loyalties and the quick rise of new transnational allegiances in the colonial sphere. See, for example, Schnurmann (1998).
18. The appropriation of local knowledge by European scholars in the sixteenth and seventeenth centuries was much more effective outside the museum sphere. See Grove (1991) and Mundy (1996).
19. Although the Royal Society officially bore the proud motto *nullius in verba*, the fellows quickly developed a far more practical attitude to travel literature (see Carey, 1997).
20. Drawing on Latour's actor-network-theory, Harris (1998) has identified the communication networks of early modern long-distance corporations as effective chains of transformation. Although his focus succeeds at broadening the understanding of the locales and actors of early modern science, the experience of collectors suggests that these networks nevertheless remained limited in scope, effectively furnishing information only if the corporation's immediate (usually commercial) interests were concerned. For Latour's model, see now Latour (2005).

References

Birch, T. (1756–1757). *The history of the Royal Society of London for improving of natural knowledge, from its first rise: In which the most considerable of those papers communicated to the Society, which have hitherto not been published, are inserted in their proper order; As a supplement to the Philosophical Transactions* (Vol. 4). London: Millar.

Burke, P. (2000). *A social history of knowledge: From Gutenberg to Diderot*. Cambridge: Polity.

Carey, D. (1997). Compiling nature's history: Travellers and travel narratives in the early Royal Society. *Annals of Science, 54*, 269–292.

Carey, D. (2003). The political economy of poison: The Kingdom of Makassar and the early Royal Society. *Renaissance Studies, 17*, 517–543.

Collet, D. (2007). *Die Welt in der Stube. Begegnungen mit Außereuropa in Kunstkammer der Frühen Neuzeit* [The world at home: Cross-cultural encounters in early modern museums]. Göttingen, Germany: Vandenhoeck & Ruprecht.

Colsoni, F. C. (1699). *Le Guide De Londres. Dedié aux Voyageur Etrangers. Il apprend tout ce qu'il y a de plus curieux, notable & utile dans la Ville* [The guide to London: Tips for foreign travelers, with everything there is to know about the most curious, notable, and practical aspects of the city.]. Den Haag: n.p.

Findlen, P., & Smith, P. (Eds.). (2002). *Merchants & marvels: Commerce, science, and art in early modern Europe*. New York: Routledge.

Granovetter, M. (1973). The strength of weak ties. *American Journal of Sociology, 78*, 1360–1380.

Grew, N. (1686). *Musaeum Regalis Societatis: Or, a catalogue and description of the natural and artificial rarities belonging to the Royal Society and preserved at Gresham Colledge/Made by Nehemiah Grew M. D. Fellow of the Royal Society and the Colledge of Physitians. Whereunto is subjoyned the comparative anatomy of stomachs and guts. By the same Author*. London: Malthus.

Grove, R. (1991). The transfer of botanical knowledge between Asia and Europe, 1498–1800. *Journal of the Japan-Netherlands Institute, 3*, 160–176.

Hall, A. R., & Hall, M. B. (Eds.). (1965–1986). *The correspondence of Henry Oldenburg* (Vol. 13). Madison: University of Wisconsin Press.

Harris, S. J. (1998). Long-distance corporations, big sciences and the geography of knowledge. *Configurations, 6*, 269–304.

Hatton, E. (1707). *A new view of London, or an ample account of that city* (Vol. 2). London: Chiswell.

Hunter, M. (1989). *Establishing the new science: The experience of the Early Royal Society*. Woodbridge, England: Boydell.

Inquiries for Suratte, and other parts of the East Indies. (1666). *Philosophical Transactions of the Royal Society, 2*, 415–422.

Latour, B. (1999). *Pandora's hope: Essay on the reality of science studies*. Cambridge, MA: Harvard University Press.

Latour, B. (2005). *Reassembling the social: An introduction to actor-network-theory*. Oxford, England: Oxford University Press.

Livingstone, D. N. (2003). *Putting science in its place: Geographies of scientific knowledge*. Chicago: Chicago University Press.

Lux, D., & Cook, H. (1998). Closed circles or open networks? Communicating at a distance during the scientific revolution. *History of Science, 36*, 179–211.

Moser, B. (2001). Neues Wissen für Europa. Die Korrespondenz und das Museum des Jesuiten Athanasius Kircher [New knowledge for Europe: The correspondence and the museum of the Jesuit Athanasius Kircher]. In R. Wendt (Ed.), *Sammeln, Vernetzen, Auswerten. Missionare und ihr Beitrag zum Wandel europäischer Weltsicht* (pp. 45–74). Tübingen, Germany: Narr.

Mundy, B. (1996). *The mapping of New Spain: Indigenous cartography and the maps of the Relaciones Geográficas*. Chicago: University of Chicago Press.

Piso, W. (1658). *De Indiae utriusque re naturali et medica libris XIV* [Fourteen books on the natural and medical things of both Indies]. Amsterdam: Elzevir.

Schaffer, S., & Shapin, S. (1985). *Leviathan and the air-pump: Hobbes, Boyle, and the experimental life*. Princeton, NJ: University Press.

Schnurmann, C. (1998). *Atlantische Welten. Engländer und Niederländer im amerikanisch-atlantischen Raum (1648–1713)* [Atlantic worlds: The English and Dutch in the American Atlantic (1648–1713)]. Cologne: Böhlau.

Shapin, S. (1994). *The social history of truth: Civility and science in seventeenth-century England.* Chicago: University of Chicago Press.

Smith, P. H. (1994). *The business of alchemy: Science and culture in the Holy Roman Empire.* Princeton: University Press.

Sommer, M. (1999). *Sammeln. Ein philosophischer Versuch* [Collecting: A philosophical essay]. Frankfurt am Main: Suhrkamp.

Teixeira, D. M. (2004). Die Naturgeschichte Brasiliens in der Regierungszeit Johann Moritz' von Nassau-Siegen (1637–1644). Die Bücher von Georg Marggraf und Willem Piso [The natural history of Brazil during the reign of John Maurice, Prince of Nassau-Siegen (1637–1644): The books of Georg Marggraf and Willem Piso]. In G. Brunn (Ed.), *Johann Moritz von Nassau-Siegen, der Brasilianer (160–1679): Aufbruch in neue Welten* (pp. 76–89). Siegen, Germany: Waxmann.

Waller, R. (Ed.). (1705). *The posthumous works of Robert Hooke: Containing his Cutlerian lectures and other discourses, read at the meetings of the illustrious Royal Society.* London: Walford.

Is the Atrium More Important than the Lab? Designer Buildings for New Cultures of Creativity

Albena Yaneva

Responding to criticism that the pioneers of laboratory studies have neglected the architecture of science labs and have failed to consider the importance of space for scientific practices, a few authors have recently shown an interest in the design and planning process of science buildings. They have convincingly demonstrated the extent to which the power of laboratories depends on sequestrations achieved with walls and doors and have explored how architecture might challenge or compromise the cognitive authority of experimental science (Galison & Thompson, 1999; Gieryn, 1998; Murphy, 2006; Shapin, 1998). Surprised that buildings have rarely figured in sociological theories, Gieryn (2002) conducted a series of studies on science buildings to address the relationship between the quality of space and the quality and identities of science and between design principles and design process. He also sought to account for the different participants and negotiating strategies entailed in the endeavor of architectural design (see also Gieryn, 1999). Other work, too, has tackled some of the issues dealt with in these studies. Galison and Thompson (1999), for instance, have examined architecture's role in the shaping of scientific cultures and identities; Livingstone (2003), the situatedness of scientific activities; and Gieryn (2006), the importance of space for both the production of scientific knowledge and the credibility of scientific claims. These researchers have striven to enrich post hoc readings of recently completed buildings by reconstructing (through interviews and archives) the design-related decision-making process that led to their physical construction. They spoke with contemporary architects such as Robert Venturi, Scott Brown, and Mosche Safdie, raising many new questions about the connections between architecture and science and debating them on the pages of *Nature*, *Science*, and the specialized journals of science and technology studies. For example, does the architecture of science determine the science conducted inside buildings (Galison & Thompson, 1999)? Can architecture shape science? To what extent are the work environment and research tools important when trying

A. Yaneva (✉)
School of Environment and Development (SED), Manchester Architecture Research Centre (MARC), University of Manchester, Manchester M13 9PL, UK
e-mail: albena.yaneva@manchester.ac.uk

P. Meusburger et al. (eds.), *Geographies of Science*, Knowledge and Space 3, DOI 10.1007/978-90-481-8611-2_7, © Springer Science+Business Media B.V. 2010

to lure scientists to new institutions outside well-established corridors of science (Powell, 2003)? Can the physical environment really influence scientific creativity (Bonetta, 2003)?

The current attention to good lab architecture has its roots in the Salk Institute for Biological Studies (1965) in La Jolla, California, designed by Louis Kahn. Its architecture relied on the assumption that good design could be central to making the institute a success by creating a welcoming and inspiring environment. Scientific research labs have not traditionally been noted for their architecture, and even the acclaim bestowed on the Salk Institute did not prompt other "signature" architects to build laboratories. As Kemp (1998) argued: "The lab is a major building type. We have come to expect little of it—other than as providing functional spaces which almost invariably prove to be inadequate as soon as they are occupied" (p. 849). Only recently has large-scale design experimentation in leading fields of science gradually begun to change the profile of science buildings as traditional structures built to support the sciences—typically utilitarian, homogeneous, and passable as a hospital or nameless corporate office. After a long past as a synonym of humdrum functional classroom buildings that fulfill technical objectives, science buildings today, with their innovative shapes and ground-breaking architecture, compete with museums and art galleries as types of buildings at the cutting edge of design. They now usually result from collaboration between brand-name architects such as Sir Norman Foster, Rafael Viñoly, Frank Gehry, the British firm of Robert Matthew and Stirrat Johnson-Marshall (RMJM) on one side and lab consultants on the other. These modern structures are seen as a powerful mechanism inducing great science, raising the public profile of researchers, and enabling them to improve their science.

At the same time, this new marriage of architecture and science is a victim of a twofold process. First, universities compete for the best minds, so they attempt to attract an array of eminent scholars in the new and highly competitive fields, such as biotechnology, IT industries, medical research, chemistry, genome research, and nanotechnology. The rapid development of those fields is substantially restructuring research activities as new patterns of communication coalesce among the main actors in the world of science. The highly interdisciplinary research and complex tasks that are involved require scientists from different disciplines to collaborate on solutions. This change in the way science is pursued greatly influences the practices of the architects who design the necessary laboratory buildings. Second, the architecture of science buildings affords and facilitates (rather than outright determines) new ways of conducting research, developing partnerships, successfully negotiating with the business sector, and extending scientific networks.

From the Lab to the Atrium

Up to the 1950s, laboratories were regarded as traditional black-box buildings with segregated spaces. Science buildings used to impede interaction among scientists from different disciplines with protected areas and enclosures keeping researchers physically isolated. The hard, sterile work surfaces discouraged sociability, with the highly controlled mechanical environment contributing to the barriers. These types of separations were specific to an archaic model of laboratory design prevalent in the 1950s and 1960s. Most science buildings at that time were designed as big window-less boxes that consisted of inflexible laboratory modules reflecting the hierarchical nature of science. The offices were tucked away in labs, making them inaccessible and inconvenient (Collins, 1999).

By contrast, the type of "generic" laboratory pioneered by Louis Kahn's Salk Institute offered wide-open expanses to make for interactions among scientists and to reconnect spaces and flows of human and nonhuman actors. The Lewis Thomas Laboratory (1986) built by Robert Venturi in Princeton, New Jersey, con-vincingly illustrates the generic laboratory model (Galison & Thompson, 1999). Payette Associates, recognized as one of the US experts in the design of molec-ular biology laboratories, was commissioned to develop the program and design of the interior spaces, and Venturi, Scott Brown and Associates (VSBA) were commissioned to work on the façade and site plan. The use of moveable fix-tures and glass windows instead of solid walls in this building brought crowds of human and nonhuman actors to free-float from place to place, intensifying their interaction.

This design had the users of the building pass through specific places and meet on its generous staircase, an arrangement that invited exchanges between floors and caused researchers to cross paths and tell each other of their laboratory experi-ences. The visual connections in the research space sharpened awareness of fellow researchers and encouraged continuous exchange of information among investiga-tors. By making it possible to share spaces, resources, and facilities and by thus promoting knowledge-sharing effectively (Gieryn, 1999), the open lab design let new connections be shaped among scientists from different fields and let new group-ings and research communities be assembled architecturally. The design of open sharable spaces for collaborative research affected the way scientists at Princeton performed science in the 1980s and had a major influence on their scientific culture (Galison & Thompson, 1999; Gieryn, 1999).

The spatial and technical flexibility of labs has become an important feature of science buildings and a way to keep up with the rhythm of scientific research. That flexibility is ensured primarily by creating a layout that allows for the reassessment of space, incorporates a robust infrastructure, and supports a mechanical distribution that can accommodate change. The design of science buildings is conditioned by this paramount need for flexibility as research and researchers change with the pace of contemporary science.

The design of the new generation of science buildings permitted a great deal of flexibility, for they can be easily reconfigured and subdivided as needed. Discussing the design of the new Chemistry Research Lab in Oxford, for instance, RMJM architects (2004) explained that "the building concept is geared around the need for flexible generic laboratory space capable of being adapted to suit changes in the 'style of science' in the future." RMJM has undertaken a number of labora-tory projects, a line of work that has developed into a specialty within the RMJM business: "The notable issue is the need to design buildings that have longevity due to the changing nature of Science. One skill is identifying through a detailed briefing process the 'specific' areas where the space can be more 'generic'" (Chris Abell, Studio Director, RMJM, as quoted in Collins, 2006, p. 16).

Thus, most of the specialized laboratories are made up entirely of demountable systems consisting of modular lab benches, cantilever casework, and modular partitions. Whole labs can be taken apart systematically and reconfigured to fit the needs of the next research project, with the operation entailing significantly less labor, time, and disruption than is possible with conventional construction. Laboratory design can thus stay abreast of the dynamics of research projects and the speed of scientific developments. Differentiating buildings by their paces and rhythms of variation will lead to groupings that might seem unusual at first glance. A chemistry or molecular biology lab, for example, might share more family resemblance with a contemporary art gallery or an airport than is first apparent and might therefore be classified with them in the same group of buildings that are changing at a great rate of speed. By the same token, a parliament building and a shopping mall might constitute a separate category of buildings, one characterized by a slower cadence of variability. Understanding and regrouping buildings by the particular variation of pace they have in common rather than by the number of fixed programmatic or aesthetic features they share will eventually change the way people apprehend contemporary architecture.

If it was important for architects in the 1960s, 1970s, and 1980s to overcome the archaic design of research labs, their challenge today is to reinvent the spatial geometry and cognitive potential of spaces for collaborative research and networking so as to enhance the social performances of science buildings. It is thus not surprising that the architectural space most discussed, loved, disliked, praised, and contested by clients (scientific managers, sponsors, and scientists) and the architectural community (historians, critics, lab-design consultants, and architects) is no longer the lab, whether of archaic or generic type. The emphasis is now on the atrium as a specific cognitive environment conducive to networking, one that gives rise to new collaborative synergies and partnerships.

The atrium is a large open space, often several stories high, covered by transparent or translucent material and usually located immediately beyond the main entrance. In ancient Greek and Roman houses, the atrium served as a particular form of courtyard, as the social center. It has developed in history as a unique form of architecture (Bednar, 1986) with vast design possibilities for creating covered interior spaces that protect users from the climate while still enabling them to enjoy the light, view the open sky, and be part of a highly interconnected communicative environment. The atrium of contemporary science buildings combines a range of programs, fulfilling a key role in the scientific life of many research institutes. Bridging interior and exterior and employing different design approaches to create open spaces bathed in natural light, the atrium invites the creation of new types of associations among scientists from different disciplines and among scientists and nonscientists. It also leads to new alliances among disparate objects of research that never could have met without the atrium's mediation.

It is no longer a matter of whether the architecture of science is a principal factor in the creation of an interactive culture for scientists; existing buildings have already answered that question affirmatively. Going beyond a merely subjective philosophy of communication, the issue at stake is *how* the spaces for collaborative science can aid the creation of new cultures of creativity. I now explore the ways the atrium and its diversity over the last decade mediate and increase connectedness, decongest activity, trigger sociality, and improve the networking capacity of science buildings—in short, how this architectural form sets houses of science radically apart from austere and faceless campus buildings.

The Atrium as a Connecting Mechanism

The atrium is becoming an important interactive space in the new generation of science buildings, a place where scientists can meet and talk to each other. The need for this kind of space has emerged with the trend toward bringing together researchers traditionally scattered across a university campus and having them work in an interactive environment under one new roof in the hope that sparks will fly. The open staircases and atria create sight lines between floors, departments, and an assortment of activities.

One finds a superb execution of these ideas at Oxford University's Chemistry Research Laboratory (2004), or CRL, built by RMJM to the highest specifications and intended to confirm Oxford chemistry's dominant position within the scientific community (see Fig. 1). The building itself epitomizes the advancement of laboratory design and won the award of the Royal Institute of British Architects (RIBA) in 2005. The CRL affords a high-quality working environment for four types of research: organic chemistry, inorganic chemistry, physical and theoretical chemistry, and chemical biology. They all converge in a specially designed space: "A separate wing providing professorial office suites and research office space, together with meeting and seminar rooms, is linked to the main laboratory wing by a glazed atrium that provides a social focus and 'heart' for the building" (RMJM Architects, 2004).

Defined by RMJM architects as being a social core of the building, the CRL atrium has an important connecting function, that of bridging the office and research spaces. Its communal meeting space with catering facilities and furniture for social contact is meant to engender professional communication and initiate collaborative actions. The CRL has made a significant academic and visual impact on the area, drawing worldwide notice and raising the profile of the scientific work within the lab. It offers a motivating environment in which scientists can work irregular hours in a pleasant and functional space commensurate to the communicative aspects of their work. Strolling in the atrium of Oxford's CRL, one does not have the impression of being in a chemistry institute.

A second chemistry building, the Chemistry Lab (2006) designed by Mitchell/Giurgola Architects/Fletcher Thompson for Western Connecticut State University in Danbury, also makes for a cooperative learning experience. Spaces

Fig. 1 The atrium of the
Chemistry Research
Laboratory Building in
Oxford, RMJM Architects.
Courtesy of Noortje Marres

in which groups can gather are available throughout the research corridors to bring about intradepartmental interaction. Inspired by chemistry, the building's materials and finishes identify the discipline housed inside. The design shows how the architects have chosen to encourage interaction with the technological mediation of particular materials. They do not just symbolize chemistry formula, valences, and reactions but also recreate an environment informed by chemistry.

A similar approach is taken at the Max Planck Institute for Molecular Cell Biology and Genetics (2001) in Dresden, Germany. The idea is to have the very architecture of this building link 25 independent research teams, an intention that explains why the atrium occupies the full height and width of the four-story building. Heikkinen-Komonen Architects, based in Helsinki, Finland, designed the building to have suites of labs and offices radiate from the atrium and connect to its central staircase by means of concrete bridges, which the scientists call *Ponte Vecchio*s, after the famous bridge in the Italian city of Florence.

Another open atrium space that furthers the intellectual cross-pollination between the fields of molecular biology, chemistry, physics, computer science, and theory is the atrium at the Genomic Institute at Princeton University (2003) in New Jersey. Using a triangular site on the south campus to its best advantage, the Viñoly architects who designed it used two perpendicular rectangular building blocks and a curved glass façade that encloses a two-story atrium space facing due south. This southern façade being transparent, the atrium space becomes an extension of the exterior. It contains a small auditorium below a circular lounge space, a number of sitting areas, and a coffee shop. A casual meeting room, in which stands a Frank Gehry sculpture originally planned for the Peter Lewis House, invites scientists to interact and exchange ideas. The changing light throughout the day and the shadows created by the lattice-like screen of louvers make the atrium a cheerful, appealing space.

The Atrium as a Mixing Chamber

Whereas the CRL atrium aims at mixing scholars from the different branches of chemistry in one core discipline, the atrium in other research buildings successfully goes beyond the simple function of furnishing spaces for social and intellectual encounters. For example, the Stata Center for Computer, Information, and Intelligence Sciences (2004) at the Massachusetts Institute of Technology (MIT) in Cambridge, designed by Frank Gehry, challenges the conventional ideas of campus lab space. The building fosters interactions between scientists from the Computer Science and Artificial Intelligence Laboratory, the Laboratory for Information and Decision Systems, and the Department of Linguistics and Philosophy. Its flexible design and its atrium enable molecular biologists, physicists, chemists, mathematicians, and computer scientists to bump into one another along the wavy corridors of the structure, in its sole dining area, and in glass-walled meeting rooms between the lab benches and offices. It also helps mix the flows of heterogeneous participants in the complex world of science and shapes additional attachments that lead to the creation of hybrid scientific fora of human and nonhuman actors.

Scientists working in the buildings reviewed thus far in this chapter often leave the lab with a draft of an article, a sample, a laboratory protocol, and sometimes even an instrument. If not under specific restrictions when outside the enclosed environment of laboratories, these objects can circulate in diverse orbits and enter other networks in which sporadic interactions among scientists may make them part of a collective of rather complex nature. All these objects to which researchers feel attached and with which they share laboratory life (Latour & Woolgar, 1979) make the world of science exotic, a world that others cannot apprehend without understanding the very nature of these attachments. And that is why erecting strict barriers between human and nonhuman actors would contradict the logic of this world and its overall cosmology. The atrium design, as in the Stata Center by Gehry, has the means for intensifying the productive encounters of researchers attached to sundry nonhuman entities (e.g., electrons, crystals, X-rays, and mice) and for having the scientists fertilize one another's research in every possible way.

Yet creating a building that increases interaction between researchers is not enough on its own. Scientists also require spaces and tools to assist them in bouncing ideas off each other and consolidating them. With this need in mind, architects also design special areas, such as the informal lounges by Venturi in the Lewis Thomas Laboratory. They are equipped with whiteboards and different kinds of furniture that give researchers comfortable discussion space away from the lab bench. These areas inspire informal dialogue over a cup of coffee by broadening the perspective their users have on what they are studying, clarifying their focus, and steadying their concentration. If a wavy corridor improves the likelihood that colleagues will run into each other by chance, then the atrium and its new materiality creates cognitive conditions for new ontological mixtures of all the constituents of the scientific venture, including texts, and sometimes even "inscription devices" (Latour & Woolgar, 1979, p. 75) of transportable nature.

The Atrium as an Urban Knot

The historian of architecture James Ackerman from Harvard University, proclaimed the Vontz Center (2000), designed by Gehry for the University of Cincinnati, to be "the most successful laboratory, architecturally, since the Salk" (as quoted in Cohen, 2000, p. 212). Another laboratory building comparable to the Salk Institute in terms of innovative design solutions for collaborative research is the Janelia Farm research campus (2006) of the Howard Hughes Medical Institute (HHMI) in Ashburn, Virginia. It was conceived by Viñoly architects, who have been designing laboratories for a decade, including a nanotechnology lab at UCLA (2006), a genomics institute at Princeton (2003), and a neuroscience building for the National Institute of Health in Maryland (2004). Both the Vontz Center and the Janelia Farm research campus feature the inventive potential of an atrium performing the new function of serving as the complex knot of a quasi-urban network.

This commonality is anything but coincidental. Assuming that most scientific breakthroughs start not in labs or in formal meetings but rather in casual interaction among colleagues, Gerald Rubin (the director of the Janelia Farm campus), Bob McGhee (a lab design expert), and Viñoly traced connections between the physical structures of existing labs and their scientific and technological achievements. They theorized that the attractiveness and natural setting of a site, along with open, light-filled laboratories and comfortable meeting spaces, can allow scientists to get on with their work and still be well connected with their colleagues and the outside world. A successful laboratory design, they concluded, should have spaces for casual encounters, with the atrium having the main role.

Horizontally, the Janelia Farm building therefore has a layout divided into three sinuous ribbons that are bisected by two atria, each anchored by a grand cantilevered staircase giving both access to all three levels of the building and egress to the meadow. Its generously sized landing affords an engaging space for interaction. Personal comfort is a priority, and care is taken to maintain access to natural light and external views in order to reconnect scientists to the external environment.

The cultural objectives of Janelia Farm dictated an unusual design for the laboratories and support facilities aimed at achieving unscheduled interactions, collaboration, and flexibility.

> The result is a major, yet transparent building with over a mile of structural glass walls. (Rubin, 2006, p. 212)

The institute's mission is to conduct biomedical research, and it has two areas of scientific emphasis. The first one, which centers on the use of genetic model systems in conjunction with imaging, electrophysiological, and computational methods, is to identify the general principles that govern how information is processed by neuronal circuits. The second main task is to develop imaging technologies. Beyond the purely scientific objectives, the Janelia Farm lab is designed to

> provide a select group of scientists with the facilities, finances, and freedom they need to pursue original, long-term research with minimal distractions. The hope is that they will, together, form a community of scientists with varied expertise and an interest in interdisciplinary research. HHMI will fully fund their work, encourage their personal hands-on research, promote inter-group dialogue and collaboration, and provide abundant research support services. (Rubin, 2006, p. 209)

To generate such a community of scientists, the lab organization is reminiscent of a city (Bonetta, 2003). It has research and conference facilities, hotel accommodations, apartment housing, a dining room, a fitness center, and even a campus pub. The Janelia Farm lab offers space, infrastructure, housing, and research funding so that visiting scientists from around the world can come together and solve interdisciplinary problems. All aspects of this technologically sophisticated research center, especially the design of the building and its infrastructure, stimulate the multidisciplinary, team-driven research needed to advance medical science. Because the problems confronting medical research today are too complex to be solved in a traditional research environment, scientists working on them require settings cognitively tailored to the elaborate research tasks entailed. The urban character of the architecture of science is intended to meet those demands by increasing research mobility, enabling research groups to relocate quickly, and allowing them to work together on a problem-solving principle for up to several years in a project-based rhythm.

To keep the scientists from getting lost, especially in huge interdisciplinary centers employing hundreds of researchers, architects like Viñoly and Gehry are thus modeling labs on real cities. These layouts have "street" areas for interaction and "neighborhoods" or large urban settlements where researchers can retreat into relatively small groups or become part of large-scale flows, respectively. Describing the Stata building, Gehry says: "The building as a whole works rather like a city: research laboratories are located within the tower wings and are divided into neighborhoods" (Rappolt & Violette, 2004, p. 342). This design philosophy does not rely simply on maximizing the chance encounters between scientists to spark new directions in research. The underlying assumption is that increased circulation within the spaces will bring scientists to experience a sense of community within a large building or campus. This feeling is expected to raise the scientists' incentive to move about in the building along its spatial transitions and various circuits. The idea is both to isolate the scholars in the highly contained environments created by research departments and the specific materiality of their science and to get them to congregate despite the segregation of their academic disciplines. In such complex city-shaped science buildings, the atrium gains the status of an important urban knot of the building network. It becomes a nexus concentrating and redistributing flows of activities, isolating and reconnecting scientists, making human and nonhuman actors circulate with greater intensity. The atrium maps their locations, guides their movements, and mediates the transactions between them. It thereby helps science buildings do something more than merely host or give rise to sporadic intersubjective encounters. It triggers new collaborations, catalyzes ontological mixtures of human and nonhuman actors, and nurtures both interdisciplinary networking and new forms of scientific partnerships.

Architects in Search of New Cosmologies of Science

What has revolutionized the architecture of science, what has changed the shape of laboratory buildings and the patterns of collaborative research in the last 2 decades, is not the merit of talented and innovating star architects but rather the "harmonious and tense, but always compelling relationship between scientists and architects" (Galison & Thompson, 1999, p. 3). The architect's creative involvement makes it possible to produce spaces for creative discovery and equips scientists with different strategic tools for positioning their fields in contemporary scientific battles over resources and best minds. By the same token, the role of scientists as users of architecture has also changed in this process and has consequently changed architecture. They have shown themselves to be engaged, well-informed clients and important decision-makers rather than remaining on the receiving end of this process.

History has sometimes shown that the architects' assumptions about what sparks creativity do not always meet the expectations of scientists as users. Louis Kahn considered natural light important for the work of scientists and designed open studios with generous natural lighting for the Richards Medical Research Laboratories (1965) at the University of Pennsylvania in Philadelphia. The scientists, however, disliked the building and attempted to modify and adjust the natural light system. Overriding Kahn's wish, they covered the lab's windows with foil so that they could continue their research. Frank Gehry intended to create a highly stimulating environment in the Stata Center at MIT, but most users of the building complained that they felt like lab rats peering out of its narrow cubbyholes.

Signature architects putting their stamp on the scientific workplace wish that their ultimate clients had greater awareness than they do of the impact a building has. Yet Gehry finds that scientists do eventually learn to appreciate buildings, especially factors like light. In this regard his experience with users is similar to Kahn's. Gehry once explained it in reference to the Vontz Center at the University of Cincinnati:

> They can turn out the lights and put a sack cloth around their heads if they want to suffer a little bit, but they are, over time, going to experience a richness. ... They will start to see how the sun falls in the atrium and how it plays with those curves. They'll start to see how the brick color was selected because at certain times of the day it has a pink glow and it's very pronounced and very interesting. They will understand that those curved walls are

nicer for a person to stand against than a big brick straight wall. So this building will unfold and have a human relationship and will enrich them. (As quoted in Cohen, 2000, p. 212)

Commenting on the design of the Stata Center at MIT, Gehry has also described the process of negotiating with the scientists on different architectures of their office spaces:

> We had a scheme based on a traditional Japanese house with panels that could open up to combine spaces and close shut for privacy. They hated that because there was no hierarchy. Then we gave them a scheme based on a colonial American house with a central hall and rooms around the bottom and rooms around the top. But they didn't like that either; it was too formal. Then one of our team members made an "orang-utan village" around a tree with elders higher up and the children below. At first they were insulted. They thought we were calling them apes. But in the end they chose the orang-utan village. (Rappolt & Violette, 2004, p. 346)

This anecdote shows how important it is for an architect to understand the cosmology of the scientists for whom he or she is designing. The architect seeks to trace the cosmogram of their scientific world by identifying and tracking the practices followed by various sets of actors (e.g., clients, actual and future users, and contractors). These actors may have mutually contradictory needs, wishes, and expectations when it comes to a new building. They may find that groups A and B are incompatible because A's research activity requires isolation and a highly controlled environment, whereas B's does not. Group A defends statement X as reconcilable with statement Y, and person C gets along with person Z but cannot share the same space with Z so would like to collaborate with colleague D, and so on. This conceptual work is quite different from simply following up on the stances related to a specific statement of agreement or disagreement with a suggested design. For this reason architects do not ask, "Are you for or against this spatial solution, this design option, this architectural idea?" All that this alternative question can do is generate different opinions. Instead, architects ask, "In which world do you live?" "How is this world structured?" "With whom and with what are you ready to share it?" "What do you cherish the most?" "Who are your allies, and who are your critics?" "How does change happen in *this* world and alter entire cosmologies?" By making some elements of the local worlds of science explicit, architects endeavor to grasp what the particular tasks and challenges of the scientific fields are, what science is, and what its cosmos is made of. They try to understand its inhabitants, what those people believe in, what they cannot live without, what they cherish the most. Design is to acquire a new role—that of progressively composing a common world of human and nonhuman actors (Latour, 1999). By prompting interconnections and mixing light, air, water flows, landscape, scientists, technicians, managers, and the tools, artifacts, and other objects of science, the atrium contributes to that meticulous work of progressive composition. By fostering the circulation of diverse heterogeneous flows of actors by means of design, contemporary architects and scientists are jointly rethinking important cosmological connections. The examples of atria spaces discussed in this chapter show persuasively that architecture does not determine scientific behavior (as argued also in Meusburger, 2008, p. 50). Science buildings rather act as mediators that make it possible to shape particular cognitive activities and to diffuse them through complex networks, to regulate flows of actors, and to actively reshape the cosmologies of the worlds of science.

References

Architects, R. M. J. M. (2004). *Chemistry Research Laboratory Building Portfolio*. Retrieved June 16, 2008, from http://rmjm.com/projects/view-bt-detail/39/Research

Bednar, M. J. (1986). *The new atrium* (McGraw-Hill Building Types Series). New York: McGraw-Hill.

Bonetta, L. (2003). Lab architecture: Do you want to work here? *Nature, 424*, 718–720.

Cohen, J. (2000). Designer labs: Architecture and creativity: Does beauty matter? *Science, 287*, 210–214.

Collins, J. (1999). The design process for the human workplace. In P. Galison & E. Thompson (Eds.), *The architecture of science* (pp. 399–413). Cambridge, MA: MIT Press.

Collins, L. (2006). Build it and they will... think? *Engineering Management Journal, 16*(4), 14–17.

Galison, P., & Thompson, E. (Eds.). (1999). *The architecture of science*. Cambridge, MA: MIT Press.

Gieryn, T. (1998). Biotechnology's private parts (and some public ones). In A. Thackray (Ed.), *Private science* (pp. 219–253). Philadelphia: University of Pennsylvania Press.

Gieryn, T. (1999). Two faces on science: Building identities for molecular biology and biotechnology. In P. Galison & E. Thompson (Eds.), *The architecture of science* (pp. 423–459). Cambridge, MA: MIT Press.

Gieryn, T. (2002). What buildings do. *Theory and Society, 31*, 35–74.

Gieryn, T. (2006). City as truth-spot: Laboratories and field-sites in urban studies. *Social Studies of Science, 36*, 5–38.

Kemp, K. (1998). Laudable labs? *Nature, 395*, 849.

Latour, B. (1999). *Politiques de la nature*. Paris: La Découverte.

Latour, B., & Woolgar, S. (1979). *Laboratory life: The social construction of scientific facts*. Beverly Hills, CA: Sage Publications.

Livingstone, D. N. (2003). *Putting science in its place: Geographies of scientific knowledge*. Chicago: Chicago University Press.

Meusburger, P. (2008). The nexus between knowledge and space. In P. Meusburger (Series Ed.) & P. Meusburger, M. Welker, & E. Wunder (Vol. Eds.), *Knowledge and space: Vol. 1. Clashes of knowledge: Orthodoxies and heterodoxies in science and religion* (pp. 35–90). Dordrecht, The Netherlands: Springer.

Murphy, M. (2006). *Sick building syndrome and the problem of uncertainty: Environmental politics, technoscience, and women workers*. Durham, NC: Duke University Press.

Powell, K. (2003). Inspiration from architecture: Building a better scientific rapport. *Nature, 424*, 858–859.

Rappolt M., & Violette, R. (Eds.). (2004). *Gehry draws*. Cambridge, MA: MIT Press.

Rubin, G. (2006). Janelia farm: An experiment in scientific culture. *Cell, 125*, 209–212.

Shapin, S. (1998). Placing the view from nowhere: Historical and sociological problems in the location of science. *Transactions of the Institute British Geographers, New Series, 23*, 5–12.

Outer Space of Science: A Video Ethnography of Reagency in Ghana

Wesley M. Shrum, Ricardo B. Duque, and Marcus Antonius Ynalvez

In May 2007, the month before the Heidelberg conference entitled "Geographies of Science," there appeared two bits of news, little-read stories beneath the notice of the regular press, apart from a few local papers. The irrelevance of the articles owed partly to their subject matter—science and technology in Africa, which is rarely reported on to begin with—and partly to their highly conventional, repetitive message of failure and promise.

The first item concerned the Nigerian earth observation satellite, NigeriaSat-1, launched in 2003 at a cost of $13 million. It had come under criticism for mismanagement. The director of mission control had been sacked. Five engineers had been accused of stealing laptops with operational software, and then fired. Six more had left, "impatient" with progress according to the head of the national space agency. The new staff was, according to others, "inexperienced and mediocre," without the British training of earlier engineers. They simply captured the data from mission centers of other satellites in the area (Jones, 2007).

The second piece of news was upbeat. At an international conference on e-learning in Nairobi, ministers and technology executives announced the exciting results of a pilot project in 120 schools in sixteen African countries. A "35% improvement in students' examination performance" was recorded, and a plan was announced to rollout the "electronic schools" initiative in 600,000 African schools within the next decade. Marketing professionals were said to be slightly worried that the project's outcomes could depend on the speed of technology adoption by teachers. But the Kenyan Minister of Education, undeterred in his enthusiasm, asserted

W.M. Shrum (✉)
Sociology Department, Louisiana State University, Baton Rouge, LA 70803, USA
e-mail: shrum@lsu.edu

R.B. Duque
Department of Sociology, Tulane University, New Orleans, LA 70118, USA
e-mail: rduque@tulane.edu

M.A. Ynalvez
Department of Behavioral, Applied Sciences and Criminal Justice (BASCJ),
Texas A&M International University, Laredo, TX 78041, USA
e-mail: mynalvez@tamiu.edu

P. Meusburger et al. (eds.), *Geographies of Science*, Knowledge and Space 3,
DOI 10.1007/978-90-481-8611-2_8, © Springer Science+Business Media B.V. 2010

that e-learning would help Kenya mitigate another problem, the need for teachers (Abwai, 2007).

Failure and promise, mismanagement and initiative—the twin themes of African "development" since independence. Epitomized by the successfully launched but operationally problematic Nigerian satellite and by the "tested" rollout of the electronic schools program, they account for this chapter's concentration on an "outer space of science," one of many in Africa, Latin America, and Asia. The particular space is Ghana; the methodology is video ethnography; and the analytical perspective is based on the concept of reagency, "a process of redirection involving a contingent reaction between identities" (Shrum, 2005, p. 723). Although the story told in this chapter is usually categorized as one of development, that designation is misleading, if not destructive to understanding. In what follows, we describe our alternative concept, reagency, and explore the argument that distance lends autonomy. We do so through a video ethnography entitled *Outer Space of Science*, which we produced for the conference. The movie documents some of the efforts that were part of a project to facilitate the Internet connectivity of two research institutions in Ghana. The central themes are spatial and temporal relationships between bodies, new technologies of communications, and the social networks heard of and encountered throughout the film. Our discussion also addresses two failings of the movie and two of its characters, one of whom is the hero.

A Space for Reagency

In the new millennium, it is preferable to consider the globalization of science by replacing the concept of development with that of reagency. As defined above, the latter highlights the social interactions and processes—including those having to do with new information and communications technologies (ICTs)—that are set in motion by streams of resource flows from "developed" to "developing" areas of the globe. The concept of reagency depends on the crucial notions of identity and place.

The central interactions in this micro-oriented perspective are between Hosts (persons whose time is primarily spent "in place") and Guests (persons who "come from afar"). We note that guest–host interactions between "developed" and "developing" areas are relatively recent in human history. Even when limited to the academic Guests and their African Hosts, the frequency of Guest–Host interaction prior to the past century does not merit attention. As Jöns (2008) has shown in her study of African visits by Cambridge faculty, there were only three trips *altogether* between 1885 and 1924. During the entire period that she studied (1885–1954), neither of the two premier British institutions of higher learning made even a single recorded visit to Ghana, the "outer space" of our movie. This embeddedness of the identities of Guests and Hosts in places structured by resource inequities is what gives their interactions a reagentive character, like that of a substance used to produce a chemical reaction. Because this book is concerned with science, the Guests and Hosts of primary concern are researchers and educators, employees of government agencies, NGOs, and universities.

Reagency is antidevelopmental in two senses. One is a skepticism about the hypothesis that initiatives, programs, and projects that come from afar have broadly beneficial outcomes. As indicated above, the other sense is the conceptual point that "development" does not well describe processes at any level of analysis except the institutional. (There are, after all, development organizations and initiatives.) The conference in Heidelberg provided an opportunity to take stock of the African wing of our project, to check data and analyze footage from a 2-year period of work. This project was conceived of as an investigation in the sociology of science with a focus on the analysis of social networks. Yet it quickly became engaged in providing funds and equipment for improving Internet connectivity. Our general argument is that distance lends autonomy to actors, reducing the "power" of the core and undermining the received wisdom that greater resources somehow compel directed action. Indeed, if it were true that power and resources mattered, then either conspiratorial forces that prevent development would exist or whatever is meant by "development" would already have occurred to some significant degree.

But neither of these conditions holds. The concept of reagency, as applied to the scientific sector, points to scientific institutes as physical installations whose irreducibly "stationary" character allows us to elaborate the theory of the spatial concentration of knowledge (Meusburger, 2000) beyond the distinction between center and periphery. We are skeptical that the notion of a global hierarchy of centers captures what is important about science in Africa. As we shift to the local and micro frame of reference, an emphasis on the spatial rootedness of social systems teaches that identities located at scientific "centers" such as the United States and Europe may reposition themselves to new organizations in Africa, but they exercise no more influence over events at these centers than any other co-located agents. A relational approach to understanding "center and periphery" represents situations of interaction as the fundamental stratum of those far-flung interactions that are called "global." Centers are locations of rather frequent pilgrimage.

The Movie

Outer Space of Science was designed as a two-part film representing 2 years of progress, or, better, events during our work on African connectivity in universities and research institutes in Ghana and Kenya. We decided to work with three primary themes reflecting the objectives of the workshop as expressed in position papers and abstracts of other presentations. The first theme was the ways that spatial aspects have been shaping inquiries into both the production and circulation of science (Livingstone, in this volume). The second theme was the idea that new telecommunication technologies have not made the distinction between center and periphery obsolete (Meusburger, 2000). The third was the observation that social networks, particularly those structured by visitations, have been crucial to understanding the dynamics of research collaboration and data acquisition since the nineteenth century (Jöns, 2008). Spatial questions were to be crucial throughout the film because the positioning of bodies in space could be shown to have an impact on the process of providing equipment to institutions, on the project objective of facilitating Internet connectivity, and on the means of collecting data on the scientists' Internet usage. That very connectivity—the new ICTs that offer the promise of reducing the impacts of distance—was to be cast as our intellectual problem during this decade. The two primary research problems were (a) the manner in which those technologies were used by scientists, educators, and researchers; and (b) the consequences of that usage. Would distinctions between center and periphery continue to hold with the reduction of the time it takes to communicate globally and of the attendant potential for remote collaboration? How would these new ICTs affect the social networks of scientists?

During our early work in Ghana, India, and Kenya, it became apparent that we would need to work with the local research organizations and use some of our project funding from the US National Science Foundation to help them establish and maintain Internet connectivity. After all, there was little point in studying the impact of ICTs in institutions where there were none, and where scientists, if they sent an e-mail at all, were simply going to Internet cafés. Because our reagency perspective was oriented to interactions at the microlevel that results from the organizational initiatives that import programs and projects into distant lands through agents from "developed" countries, we became objects of our own theory. With complete access to our own discussions and those of our permanent collaborators and temporary visitors, we began to develop video ethnographic techniques to follow the connectivity initiative in Africa and India, through the National Science Foundation program on Information Technology Research, to our own small project in three sites and six institutions in Ghana, Kenya, and Kerala (India).

By 2007 we had been collecting footage for approximately 5 years. From 2002 through 2004—the most active period of filming our attempts to make a difference to research institutes and university faculties—we reviewed 70 h of footage shot in Kenya and Ghana during the summer research visits for a selection of scenes illustrating various aspects of reagency. Although the editor (Shrum) was already familiar with the events (as a participant) and the footage (as either a cameraman or a subject), four undergraduate students reviewed each hour of footage twice. On the first pass, the objective was to become familiar with the material and to identify the segments that were audio-friendly and pertinent to the story line. On the second pass, attention shifted to producing specific and detailed summaries of important episodes. Although the footage and the stories developed differently in Kenya and Ghana, several individuals in Accra (Ghana) seemed to provide a kind of linear narrative through the vicissitudes of video ethnographic work. Because our fundamental arguments over what we were doing and why we were doing it were beginning in earnest during that period, it seemed best to focus on the West African events in order to weave a generally linear narrative.

There were two primary flaws in the Heidelberg presentation, as indicated in the discussion at the conference. One was the understanding of the narrative voice; the second was the identity of the two principal characters in the film, one seen frequently and one never seen (the names of some people referred to in the film were altered for purposes of confidentiality). We deal with these characters below and again in the conclusion. To address the first shortcoming as we examine the method of video ethnography, certain presentational aspects warrant scrutiny. An audio track in a movie (analogous to the talking of a scholar behind the podium during a presentation) can take one of two forms: "voice-over" or "in situ." In a voice-over, the author of the movie provides verbal commentary as an audio context, background, or interpretation for what the audience sees on screen (the video track of the movie). This contribution occurs after the fact and from a different place, usually from an editing suite or office. A voice-over is retrospective, just like the interpretive net a traditional scholar casts over collected field notes or archival materials. In the conventional voice-over, the author has complete control over the ex post facto elucidation of an event. But what appears to be voice-over in *Outer Space of Science* is not of this nature; it is *emplaced*, temporally embedded interpretation. The voice that is heard was recorded at the same time as the associated video scenes. Of course, there is always a choice made to utilize such interpretation in the final analysis, for the selection process cannot be abolished. But this method dramatically reduces the scope for selection—audio tracks that are derived from original footage can only be "cleaned," not significantly altered. That is, one may eliminate unwanted sounds, but one cannot "piece together" sentences and still keep on the proper side of the ethical divide.

A clear example of this technique comes near the beginning of *Outer Space of Science* when a long shot (50 s) shows a pump as the author ruminates on the analogy between a water pump and the Internet, the subject of his own development project:

> The demonstration pump, for "practical irrigation." A pump pedal, getting the water from one place and out onto the ground where it is going to fertilize [*sic*] something, some crops. We're trying to connect people to the Internet. We want to get the information out of the pipes.

This real-time monologue is marked by mistakes (*fertilize* instead of *irrigate*), stumbles, and even statements of purpose that are called into question by the subsequent story line. But the importance of this emplaced narrator is that he is on location, positioned in time and space as part of this story. Because placement of narration is far from obvious to a viewer but crucial to the understanding of an argument in an academic movie, the spatiotemporal relationship between the locus and the substance of the narration must be considered a problem that needs a solution: What new "convention" could we use to indicate whether an audio track is "emplaced" interpretation or "supplemental" interpretation in the manner of traditional scholarship? The reason such a convention is needed is that both techniques are valuable, but the audience needs to know which is being used.

The second flaw in the Heidelberg presentation was more complex and related to the identity of the two principal characters. At the conference, both of them were viewed as ambiguous, though there was no ambiguity in the mind of the editor. Let us leave our hero until the conclusion and begin with the role of the unseen Derek so as to alleviate the concern that he existed only in the imagination of the project principals. In the film Derek is shown to *reimburse* a sum of money for work that had not been completed. This behavior is unusual in the given context and is one of the movie's main story lines. Local conditions make such a refund unlikely, if not impossible, a circumstance the viewer does not discover until near the end of the film. Our intention was to use this singular event to shed light on the process of reagency and its typical forms. Why would it be so difficult and exceptional for an individual to give back a financial advance that had been provided to start work on a contract? One reason—but the least interesting—is the near impossibility of proving that no work was accomplished and that there was no further claim to payment on completion of the work. The more illuminating reason for the anomalous act begins one year earlier, in the attempt to jump-start the connectivity project in one Ghanaian research institute and one university college.

The project personnel consisted of a director—the first author of this chapter—and several doctoral students and national coordinators in each country, each located at a major university or college near Ghana's capital city, Accra. The coordinators were locals, but they were not ICT experts. Occasionally, as depicted in the movie, volunteers would join the project members out of desire to help and to experience faraway lands, and sometimes these individuals were knowledgeable in the ICT field. In 2003 several such advisors accompanied the project director and his associate to Accra to assist in meeting with locals, developing strategies to establish Internet connectivity, and collecting data on the usage of the Internet by scientists. Speaking in a film segment after the demonstration pump has been introduced, the narrator expresses his frustration with the progress of the project thus far. He quotes the German director of a nearby international research institute:

> It must be disappointing for you. After all, if I remember, your project is to collect data and find out how people use the Internet. Seems like you're spending all your time figuring out how to get people connected. Boy, he hit the nail on the head.

The first half of the movie illustrates this condition—hitting a nail on the head may be easier than connecting an organization to the Internet, but new ICT is not unequivocally beneficial to its recipients. The first indication that it may be a mixed blessing is the experience of a white expatriate in the north of Ghana. He learns that our project is here to help with Internet connectivity and to find out how scientists communicate with the Internet. He responds, laughingly, that in some sense they would be better off without the Internet:

> "With difficulty" is how we use the Internet. I used to be up in Tamale, and that was the next best thing to nothing. You really wish there were no Internet, because people have this expectation, "Oh, send me an e-mail." It's like, "Yeah, sure, I could deliver it faster on a donke". Yeah, that's great, you know. Anything you can do for us.

These expectations are one key to understanding both the "digital divide" and every divide that is defined as worthy of resource infusions, visits—sometimes extended ones—by Guests, and their scientific programs and technological projects. Expectations for ICTs are similar in this respect, but they are also divergent in that they are the means for conducting other kinds of business. The expatriate scientist wanted improved connectivity because his needs were simple—to communicate with his colleagues on the same terms as he would in the US without the need for movement of bodies in space ("faster on a donkey").

The Guest–Host relationships that are characteristic of development projects have always involved this movement. What struck one project member just after the interaction with the expatriate scientist was that the scientist was "acting like a local" in his concern with a return or follow up visit. The theme of "return" has been pervasive since our project began in 1994 and has been noted by other scholars of the geography of science (Bauchspies, 1998). It is the expectation of return that creates relationships, friendships, and lasting commitments; but more common is its opposite, the Guest or Stranger who states an intention to return but does not. Projects that entail contracts and money generally require at least one return visit and a further visit for evaluation. Although Internet communication can be used as a substitute, almost all projects require funds for these face-to-face visits, just as they did before the advent of the Internet. Routine contact is less place dependent than such visits are (Meusburger, 2000), but the more important issues of project design and funding are subject to the "compulsion of proximity" (Boden & Molotch, 1994), which requires the movement of bodies in space, often over thousands of miles in the case of development programs.

The central interaction in the 2003 period depicted in the movie is a meeting of six individuals, three from a research institute and three from the connectivity project. One consultant questions the "demand aspect" of connectivity, suggesting that perhaps their connectivity problems do not result from issues of technological supply but rather from the question of whether "anyone cares about this enough to cause [the director] enough aggravation to make sure this is done." The director, supported by his staff, replies that his scientists go to Internet cafés if they need to send an e-mail when the Institute connection is not working. Clearly, his response does solve the problems of those who need to send these e-mails. But it is not one that the consultants find satisfactory. The project director answers that, taking everything into account, the most important issue is reliability. Low reliability teaches scientists that the Internet is not that useful and perhaps not worth learning to use.

This view is conventional but not universally accepted, as shown in the relationship between the project, the research institute, and the university. Although the decisive scenes were not subject to videotaping, the consultants do speak about their interactions with senior administrators. This meeting is clearly not the first with these individuals or their counterparts in Ghana, and the frustration of the consultants is apparent. Because reliable connectivity has not been achieved in 2003, the objectives of the project become increasingly unclear. In a simplistic way, the main interactions are characterized as negotiations over prices—the quotes are higher than they should be. In one scene, the project director asks a vendor if the prices being quoted for items are actually for *all* the items required, and the vendor responds that they are just for single items. The director is clearly aggravated.

But it is apparent at another level that the problems are more complex than getting the lowest prices for products, something that might be achieved in the US through shopping.com. Setting and getting the desired price is often a challenge in Africa, but the edginess of the interactions suggests that spatial proximity, the ebb and flow of bodies in the outer space of Ghana, is the root issue. Plans are made, agreements are concluded, and processes move in different and unwanted directions when the expatriates leave, with or without e-mail connections. At the university, the project director explains to his consultant that the given context makes it impossible to keep to a strict plan:

> I make certain gambles and invest money at certain points, hoping that I'm going to figure out the next piece of the puzzle later. Here at the university, there are certain individuals that lead me to believe more difficulties might be there than at certain other institutes and universities. I'm aware of that, so I'm going in with my eyes open. But if you ask me whether I have an actual plan that guarantees...

The project director argues that it is better to be flexible, to have a general strategy and make decisions on the fly. But the consultant, unconvinced, cuts him off: *"Guarantees* is too strong a word. Are you going to try to convince me, or are you trying to convince yourself?"

Another scene at the university was filmed just after a meeting with top administrators:

> We're hearing from the same team of individuals that has been in place since we've been coming here.... I don't think they have any interest in moving things forward. I think they have an interest in keeping things exactly the way it is [*sic*], exactly the way it has [*sic*] been for the past four years.

Aggravation has been transformed into despair:

> We've been saying the same thing: "We're going to build a connection." They've been saying the same thing: "How much money do you have?" So I don't see anything changing, and I don't see anybody getting better connectivity. And I think our final outcome was fine. We said we're going to put in a server and buy five access points, put in a few network cards, and... we'll see if anybody's connected. And if they're not, then we're not going to spend any more money. We're going to leave it like that. If that's the way they want the university to operate, there's nothing that we can do about it.

The emotional energy generated by the meeting is evident in the face and tone of the project director, who shakes his head and looks down in resignation: "These people are going to have to die, or retire... before they change."

Part of the explanation for the director's surrender resides in one individual who had become a gatekeeper. Sources of funding such as this small project once had to rely on this individual, who had no formal training in information technology. He had developed a networking plan for the university and had ties to senior administrators. Instead of providing general access to a master plan for connectivity, he kept a set of drawings stored on his laptop computer. His personal knowledge remained private, so when the price of equipment dropped and solutions became more widely available, "what he did was to monopolize the knowledge.... [H]e's trying to figure out 'how can I make money?'... This is not his job as a university servant." Throughout many sequences this suspicion, this fact of "development," lurks near the surface of interaction. Individuals who are firmly anchored in a space where resources do not originate locally are motivated to establish relationships with temporary residents who represent organizations authorized to distribute program and project funds.

> This particular place seems to have a group of people in charge of the connectivity who have figured out how to make money. As long as everybody's not connected, the money can keep coming. But once everyone is connected and it's functioning like a university in Europe, then there's no more money to be made.

The movement of these Guests into and out of their local space provides opportunities to negotiate over resources, but agreements and contracts are known to be "subject to local conditions." This movement of people and resources into an outer space is the motor of the reagency process.

An important part of our video ethnographic output is the flexibility it provides for discussion and interpretation. One of the most insightful interpretations is offered by a consultant who struggles to understand the problems faced by the project director in relation to his goals and motivations: "So your big thing is an ethical, a moral thing. . . taking your time that you could be devoting to doing something else when you're here?" More than anyone else, this person seems to relate to the ambiguities and uncertainties that produce not development but interactions of an unpredictable character, larger questions about how to proceed in spending money and signing agreements.

Director: I just want to do the right thing.
Consultant: What's the right thing. Something that feels good?
Director: I don't want to waste money. I want to give them something that will work. I want to give them something that is not outmoded.... There's just multiple considerations going on. Whether the technology is available, whether we can even get it—that's an important thing to know.

Doing the right thing is not a simple but a complex, sometimes unfathomable effort in pursuing the sociology of science. As an attempt to understand the impact of the Internet on science in developing lands, the project has become enmeshed in the problems of development, in the interaction with vendors, professors, consultants, and cablers who understand the dynamics of entering and leaving local space much better than the project director does. Generating action seems easy with face-to-face encounters: Commitments are made, and implementation is foreseeable. But frequently there is minimal follow-through—"you know how it is in Africa."

That final interaction between director and consultant makes a transition to a discussion of Derek, one of the two characters that the Heidelberg audience perceived as ambiguous in the film. He is to be hired to develop tracking software. In the second half of the movie, Derek dominates, along with the server, the piece of equipment he had been provided through the project in 2003. The consultants go back and forth, trying to find Derek, who some people say has gone to South Africa. No one seems to know where the server has gone, the implication being that it may have been stolen. The local university has been disappointed with Derek's service. The fellow may have come once or twice with his assistants. Sometimes, in the words of one local, "The boys did come, but then they said something was not in place. They were going to come back, and they never showed up."

This pattern is typical, not just of donors and expatriates but also of locals engaged in projects. Their physical presence depends on other opportunities and constraints stemming from their social networks: "It's not that anyone was trying to do anything except their job. . . .. It's that they have a lot of jobs." The director says there is only one way to find out: "We just have to go out there and see what's going on." The advisor demurs: "We send an e-mail and we say we'd like the server. If he refuses, we'll take action. We'll begin court proceedings." The statement is hilarious for the local coordinator: "Do you have lawyers in Ghana," he laughs. "We'll find some!" This exchange is indicative of the problem of accountability. Ultimately, expatriates have little recourse in the event of equipment that is purchased but uninstalled or of work that is paid but not performed. Everyone knows that Ghanaians cannot begin projects without some payment in advance, but often the advance is not enough to ensure completion.

But Derek is an exception. At the end of the movie, to everyone's surprise, he does return his advance. Why? The difference lies in the social networks that are activated within a few days after the arrival of project personnel in 2004. "We know people that he knows," says the project advisor. They are not just friends in common, but *expatriate* friends. Derek, it seems, was raised in Ghana, but went to college in a developed country at a prestigious institution. His self-image is tied, more than for most Ghanaians, to his feeling that he is, or would like to be, part of the expatriate social network. If Derek could be contacted, there would be some chance that something could be worked out. The remaining problem is how to reach Derek.

The rest of the movie deals with the search for Derek and the interactions using Internet and mobile technologies. First, at the busy Internet café, the advisor rapidly sends a series of e-mails to his and Derek's mutual friends. Until these ties are mentioned, Derek does not respond. Eventually, they meet, and Derek decides to return the advance payment rather than complete the work. Near the end of the movie, after the money has been returned, the director and the advisor argue about motivations and reasons that may have been peculiar to the case. They have divergent assessments, but the explanation of events lies in constraints imposed by timing and spatial positionings. The chain of events began when it was not possible to obtain bids for equipment before the project left Ghana in 2003. Had that happened, direct payment for equipment could have occurred. Instead, months passed before contracts could be signed and money transferred to Ghana. By the time the director and the advisor had imported equipment from Germany and were ready to have it installed, Derek had gone on to a job in South Africa. His network of contacts had provided him this other opportunity: "It wasn't his fault," says the director. "He's not supposed to wait around to earn [a mere] $500." Derek returned the money because his reputation *in the eyes of expatriates* was important to him and because the strategic use of available ICT made it possible to establish that this reputation was at stake. However, none of these circumstances would have mattered if not for the physical presence of project staff. These three themes—movements in and out of local space, network ties, and ICT—will generally be required for an accounting of reagency, but they dovetail here in a singularity.

Conclusion

Structural inequality has persisted at a global level throughout the post-World War II era of development and into the modern period of globalization. Macro approaches to development aid show that it has changed the economic relationships in the world system very little. The rise of labor-intensive manufacturing in noncore zones has led to upward mobility for only a few countries (Mahutga, 2006). Our approach occupies the opposite end of the spectrum, combining video ethnographic methods with a microsociological perspective on aid. The association of the demonstration pump and the Internet places the action of *Outer Space of Science* within the context of the problems posed by development. Something is being "demonstrated," presumably something that "works." Conventionally, science in outer spaces is developed through technology transfers, with technologies being provided by agents of development who possess greater resources than do the recipients of development. What our approach contributes is the insight that the resource imbalance is only *financial* in nature. The distance of the recipients from developed areas lends them an important kind of autonomy. They possess the important resource of locality, and they use ICTs creatively to achieve diverse goals. In one sense the story we tell is a familiar one of constraint, failure, and corruption, just like the Nigerian earth observation satellite that was launched from and into "outer space." The reagency concept conveyed through this story is not judgmental, so we finish here with the positive story of our leading character.

He was connected above with the second flaw in the movie—confusion about the identity and even the valuation of the main character. The gentleman who often appears in a bright red shirt and cap, Dan-Bright Dzorgbo, is the "hero" of the film. He is a lecturer in the sociology department at the University of Ghana (in Legon). He has been the national coordinator of our project since 2000, but more important, he has been a good friend. That friendship is based on nothing more or less complicated than frequent and enjoyable interpersonal interactions. It explains our long-term association better than any notions of trust or contractual arrangements. It is a commitment built through repeated visits over time, a pledge to engage in annual visits for the duration of the project and beyond, a mutual understanding that "something" will keep moving, even if not the something originally designed.

Dan-Bright originally hails, like most Ghanaians, from a village. His Ph.D. in Sweden and his career in academia are hard won and unique within his family, meaning for him higher income and status than most Ghanaians have, yet also the obligations they entail. In the movie he appears several times in connection with serious problems with contractor relations. But throughout he is also shown in a pickup truck, collecting fellow villagers who need a ride. At the end he is seen making plans in a peaceful, rural area. This latter sequence—totaling no more than

2 min—is woven into the fabric of the project, but the images are green and red, clear and bright. They feature the rural landscape as opposed to the brown urban institute. The movement of bodies in space is communicated by the rush of landscape across the background and foreground and by Dan-Bright's peaceful stroll through the bush after arrival. Instead of the darkness and harsh light of the hotel, the research institute, and the Internet café, instead of the frustration, confusion, and conflict of development, one sees the evident anticipation of someone returning home: "To my village" is the sequence's only English phrase alluding to these first travels.

Midway through the film the audience witnesses the "pickup," as Dan-Bright stops his truck for five fellow villagers on foot. For viewers, the ambiguity of his role as the university professor returning home may be awkward in its status implications ("I will stop for them, but they will be behind. That's what they do. They will be at the back."). Dan-Bright does not speak with them; he merely stops, explaining the action to the expatriates in front. The remaining clips in the sequence occur at a special place on the planet. For Dan-Bright, it is his home or, more rigorously, it is a space near his home. He is showing his friends "where I've just acquired a plot of land to keep up a house in my village." After his return from doctoral studies, he consulted with community elders, who gave him the plot where he intends to build a house. It is on a cliff near the junior secondary school—a symbolic location—and will afford a grand view across a valley to the village: "When I build, I'm going to see the whole village. I think that is exciting for me to be able to have the whole village in perspective." He makes the point again later, "When you are in the house, you can have a bird's-eye view of the village."

From that perch, Dan-Bright's family will be above and apart from the village, yet the joy in his eyes is apparent at the thought of moving back home. In the final scene of the movie there is dancing, now to the village rhythm that originally accompanied Dan-Bright's homecoming. Though unstated, the dancing takes place at a funeral. The author of the film—the project director—is seen for one third of a second, smiling for perhaps the first time, watching at the edge of these festivities. The last edit returns to Dan-Bright, still dreaming of the house as he looks into the distance: "When I build, I'm going to have the whole town in front of me, so it's going to be very beautiful." Like Dan-Bright, the author is joyous but apart from the villagers. In a real sense, Dan-Bright may (or may not) be going home, but the author has helped him along the way. There is little question that, though the Internet connectivity was long in coming, if it ever did come at all, the project funds helped to build a family home through a legitimate professional relationship that became a friendship. That friendship was a structure in space that came from interpersonal connectivity and not Internet demonstrations.

Because this essay and movie have described processes of reagency that are commonly viewed as malfeasance, we emphasize that understanding geographies of science, particularly the outer space of science, requires one to suspend judgment about the eventual outcomes of these processes no matter how one participates in the present. In our own project, which continues (albeit without any direct concern for connectivity), we will continue to abide by the rules we have set. We will try to prevent project funds from being misused—and we will call it stealing, should that be the case.[1]

Note

1. It is instructive to consider the East India Company's trading network, which was developed over more than two centuries, well before the current era of globalization. Erikson and Bearman's (2006) study of 4,572 voyages, based on the ship logs, journals, ledgers, and reports of English traders, shows that dense, integrated global trade networks gradually developed as an unintended byproduct of "systematic individual malfeasance" (p. 195). As ship captains sought private profit through the Eastern trade, they created fertile conditions for globalization and the geographies of science that characterize the modern world.

References

Abwai, K. (2007, June 1). *African children do better with digital.* Retrieved April 25, 2009, from http://www.scidev.net/en/news/african-children-with-digital.html

Bauchspies, W. (1998). Science as stranger and the worship of the Word. *Knowledge and Society, 11*, 189–211.

Boden, D., & Molotch, H. (1994). The compulsion of proximity. In D. Boden & R. Friedland (Eds.), *NowHere: Space, time and modernity* (pp. 257–286). Berkeley: University of California Press.

Erikson, E., & Bearman, P. (2006). Malfeasance and the foundations for global trade: The structure of English trade in the East Indies, 1601–1833. *American Journal of Sociology, 112*, 195–230.

Jones, F. (2007, May 29). *Nigerian satellite dogged by "mismanagement."* Retrieved April 25, 2009, from   http://www.scidev.net/en/news/nigerian-satellite-dogged-by-mismanagement.html

Jöns, H. (2008). Academic travel from Cambridge University and the formation of centres of knowledge, 1885–1954. *Journal of Historical Geography, 34*, 338–362.

Mahutga, M. C. (2006). The persistence of structural inequality? A network analysis of international trade, 1965–2000. *Social Forces, 84*, 1863–1885.

Meusburger, P. (2000). The spatial concentration of knowledge: Some theoretical considerations. *Erdkunde, 54*, 352–364.

Shrum, W. (2005). Reagency of the Internet, or, How I became a guest for science. *Social Studies of Science, 35*, 723–754.

The Making of Geographies of Knowledge at World's Fairs: Morocco at Expo 2000 in Hanover

Alexa Färber

> *I am often asked about the future of world fairs and whether they
> have become cultural dinosaurs. I do not have a crystal ball, but
> when I gaze into the Crystal Palace—and, just as important,
> gaze through it into the broader political and economic
> conditions that led particular individuals to build it—it is only
> the rash prophet who thinks world fairs have lost their influence.*
>
> R. W. Rydell (2006, p. 149)

The influence of world's fairs today is hard to define—their cultural resistance,
however, is evident. The viability of these erstwhile protagonists in the aesthetic
dramatization of the modern narrative is surprising because one expects new representational formats to respond to the rather moderate story lines of late modernity.
In previous ages the world's fairs of modernity "were not just exhibitions of the
world, but the ordering of the world itself as an endless exhibition" (Mitchell, 1989,
p. 218). In light of that enormous impact, what representations of today's world do
contemporary world's fairs provide?

For the past 150 years, cities have jockeyed to host world exhibitions, lobbying
locally, then nationally, then internationally, and almost always despite foreseeable
fiscal disaster. City officials defend this economically irrational behavior by playing
up the symbolic benefits to the local community and national standing. Even today,
world exhibitions attract legions of visitors over 5–6 months. They are people who
feel committed to classical national representation and have fun in the entertainment
areas, likening their experiences to those they have had as tourists and consumers of
leisure and amusement parks.

This cultural resistance, in all its historical depth and possibly in its contemporary
social relevance as well, becomes apparent in the design of the different countries'
pavilions at the world's fairs. These structures incorporate what I refer to as representational work: those often multidisciplinary professional practices that lead to
representation, in this case the representation of a nation in the form of a stand

A. Färber (✉)
Institut für Europäische Ethnologie, Berlin 10117, Germany
e-mail: alexa.faerber@rz.hu-berlin.de

P. Meusburger et al. (eds.), *Geographies of Science*, Knowledge and Space 3,
DOI 10.1007/978-90-481-8611-2_9, © Springer Science+Business Media B.V. 2010

or pavilion (Färber, 2006, pp. 15–17). Ensuing from reflective knowledge-based work, it captures the ambivalent contemporary spirit that has imbued world's fairs in recent years. The analysis of representational work that contributed to one of the latest world's fairs, Expo 2000 in Hanover, Germany, shows the ways in which carefully construed geographies of knowledge were inscribed in moderate late-modern narratives, that is, in the guise of modernity.

This chapter centers on the representational work that was invested in preparing the participation of a particular country, Morocco, and draws on analysis of the exhibit as well as on fieldwork in the Rabat office responsible for the Moroccan contribution. The purpose is to demonstrate how a small multidisciplinary group of scholars, civil servants, and architects shaped what became "Morocco: The Roots of the Future" (Generalkommisariat, 2000b) at Expo 2000 in Hanover. I maintain that much of this participation developed in anticipation of the fair as a framework structured by what was interpreted as existing geopolitical hierarchies objectified partly in knowledge divides between an ignorant public and a reflective guest country. These divides merged with a modern representational framework that was regarded as culturally plausible for visitors.

At first glance Morocco's self-representation contrasted sharply with the dominating topos at Expo 2000, the knowledge society. On the eve of the new millennium, globalization and international competitiveness were inscribed mainly in the topos of the knowledge society that had risen to its discursive climax in a variety of social spheres (politics, culture, and science). According to those who defended its promises of democratization against more differentiated (and skeptical) approaches, the knowledge society is supposed to eventuate in equal access to technology throughout the world and thereby fill in rather than reproduce the knowledge gaps that divide nations.[1]

At second glance it became obvious that the translation of "Morocco" into a recognizable representation rested on more than relational self-interpretation and self-positioning. It was also based on representational work that had the means and aesthetics of the knowledge society at its disposal, resources that the participating experts decided to disguise within the general architecture of the country's self-representation at Expo 2000. Prototypes from representational archives were reproduced and combined with theatrical elements of presentation—all classical, modern representational forms that world's fairs had once perpetuated, if not invented, as urban attractions (see Benedict, 1983; Bennett, 1988; Mitchell, 1989; Rydell, 1984). Morocco's resulting exhibition stand at Expo 2000, instilled as it was with an oriental atmosphere, was voluntarily rendered invisible as a possible knowledge society.

Face to Face: Representational Work and Knowledge Work

For our own age which is so concerned to deconstruct meanings, fairs are consummate texts: they were planned and executed by committees with conflicting agendas and contradictory purposes. By digging deep, we can discover veins of rich symbolic material and lavish deposits of meaning. (J. Gilbert 1994, p. 14)

To shed light on "the broad dreams of planners and promoters" (Gilbert, 1994, p. 13) and the realization of their projects, I focus on a case study of the representational work executed for the Moroccan pavilion at Expo 2000. My argument derives from the perspective of a guest country at Expo 2000, in this case the kingdom of Morocco. My research in Rabat and Hanover afforded insight into how the given structure, legacy, and local German version of the world's fair in 2000 was perceived, interpreted, modified, and affirmed in representational work. As an analytical term, representational work offers a perspective on the "exhibitionary complex" (Bennett, 1988) as both a performative representational format and a knowledge practice that modulates and is modulated by the given representation (in this case, the exhibit).

The representation of Morocco is particularly interesting to analyze partly because of the country's self-proclaimed intermediary situation, which is deeply rooted in the country's cultural imaginary. That self-concept has resulted from Morocco's geographical location in northern Africa on the southern shore of the Mediterranean and from the country's historical development into a multiethnic and multireligious society because of intermittent occupation, colonialism, conversion, and migration. At Expo 2000 Morocco was presented as the "bridge between" Europe and Africa, the geographical tie between north and south, the cultural transition between east and west, a land between orient and occident, between tradition and modernity. Or as publicly declared by the motto of the exhibition, Morocco was a country with "roots of the future."

A close reading of the representational work in Rabat and the exhibition tools used at the Moroccan national contribution to Expo 2000 bring the general contradictions of postcolonial self-representation to the surface. From a postcolonialist standpoint, the assertion that Morocco was compelled to envision itself as the "oriental other" at the world's fair so as to attract attention and successfully reposition itself internationally can be seen as strategic mimicry of a colonizer's (or former colonizer's) viewpoint and offers little space for reinterpretation. Representational work for the national contribution at Expo 2000 was in fact profoundly ambiguous for all countries that participated in the event. The guest countries were confronted from the very outset of their preparations by the obviously restricted local agenda of the host city. World's fairs always make it known that "their host communities were active and dynamic" (Harris, 1990, p. 127),[2] and Hanover was no exception. As the capital of Lower Saxony, it sought to strengthen its standing as a city for trade fairs, expected local development from the investment in infrastructure, and tried to boost the symbolic capital of the city. Morocco's participation in Expo 2000 was meant to do more than simply enhance Germany's image as a cosmopolitan host country. As explicitly stated in the official speeches by leading German politicians (Press Office, 2000a, pp. 6–7, 2000b), it was also to disassociate Germany from recent zenophobic aggression indirectly recalling Germany's past and to dispel other lasting traces of Germany's national socialist history (1933–1945). Back home, however, the Moroccan contribution, indeed the entire event, passed more or less unnoticed, a fact acknowledged by my interlocutors in the Rabat office in charge of preparing Morocco's contribution. After all, calculated in terms of the average number of minutes each visitor spent at any one exhibit, all countries represented at Expo

2000 faced fierce competition for on-site attention. Undeterred, though, Morocco based its participation on complex considerations of future global, international, and regional positioning and a wish to attract tourists to the country.

My fieldwork consisted of participant observation at various points in the pavilion's development, with intervals from early March through late August 1999 and intensive research on site in Rabat from September though October 1999 and at the exhibition site from June through October 2000. Within the Commissariat Général (CG)—the official Moroccan agency responsible for the country's exhibit at Expo 2000—this research approach promised insight into the representational work done within the specific professions and social milieus embodied by the Moroccan team. Who made up that strata in a society marked by vast social differences and at times barely held together only by a powerful monarchy (see Hammoudi, 1997)? To answer that question, one must first realize that Morocco has undergone important transformations since the 1980s despite its authoritative political structures. In 1998 these shifts led to *alternance* (widely understood as broad change) in governance. Abderrahmane Youssoufi, a former political prisoner and still the leader of the socialist opposition party L'Union Socialiste des Forces Populaires, was appointed as prime minister, and Islamic representatives entered parliament. The idea was to shape a political culture of dialogue, including symbolic recognition of a whole generation of oppositional politicians and of the growing impact that middle-class activities had on the construction of the country's civil society for 20 years. The social question had become a central political issue (see Khatibi, 1998).[3]

This civic involvement was clearly mirrored in the composition of the office in Rabat. The core group consisted of 5 people: the Commissaire Général (a former State Secretary in the Ministry of Agriculture and the Moroccan ambassador to Germany at that time); the ambassador's assistant (an agronomist and office manager in the Ministry of Agriculture); two accountants; and a part-time advisor on economics and international fairs. The Rabat office also had eight permanent advisors. Two of them, both trained ethnologists, were civil servants from the Ministry of Culture and were responsible for museum and cultural heritage. There were two geographers from the university in Rabat, one an expert on migration and tourism and the other on water and sustainable development. The fifth permanent consultant was a political scientist and anthropologist specializing in religious social movements. The other advisors were a musicologist, one representative of the trade association, and someone from the architect's office (either the architect himself or his assistant). Almost everybody having to do with the office's representational work had a hand in building Morocco's emerging civil society, whether in the realm of environmentalism, family and women's rights, or the representation of minorities (e.g., the Berber population). The social question fed a certain *esprit militantiste* throughout the sober rooms of the CG's members in their well-situated neighborhood in Rabat.

How did these diverse experts from various disciplines take what they understood as Morocco's reality and translate it into an exhibit presenting the country to the international community?[4] And how was this local, project-based work to be inscribed into another local interpretation (Expo 2000) of a 150-year-old representational format—the world's fair?

From the Communication Strategy to the Stipulations for Competitive Submission of Design Proposals

The overall semantic structure of Expo 2000, the motto "Humankind—Nature—Technology," clearly avoided dramatic rhetoric about progress. Instead, an unstructured, open field of resources was suggested as a central theme for all participants.[5] The CG's task was to construct a Moroccan stand consistent with the motto. From a list of ten subsidiary themes, the Moroccan representatives chose "mobility" and "sustainable development," which then had to be translated into a genuine representation of Morocco. The result at Expo 2000 in Hanover was "Morocco: The Roots of the Future."

The CG was intent on ensuring that the representation of Morocco offered dependable knowledge rather than stereotypical images:

> National participation at the universal exhibition must be based on *reliable and credible information*, measurable and verifiable in the facts and in the field. It must reflect the trump cards of the country and actual dynamics that face the challenges of the next millennium.[6] (Note de synthèse, 1999, p. 5, author's translation)

As instructed in the CG's guidelines and strategy papers, this mimetic aspiration was repeated constantly[7] and at every meeting during my fieldwork at the CG in Rabat. The risk that the exhibition site would create a gap between "fact" and representation was also discussed with all the architects, the collaborating representatives from the ministries and universities, the people in competing communication offices, and artists working on the project or planning to become involved.

The average visitor to Expo 2000 was expected to spend only a short time at each country's stand. This assumption instilled awareness of the "risks of representation" and led to the formulation of a general communication strategy: Clear and Concise Messages. To compete with the overwhelming supply of information across Expo 2000 as a whole, the Moroccan contribution was to pursue "an offensive strategy that readily *attracts*, *informs*, and *convinces* a big part of the 40 million expected visitors" (Avant projet, 1999, p. 2; italics added). Accordingly, all contributions by CG members were to address the issues of mobility and sustainable development, accommodate the specific constraints of a mass event, and convey Morocco's image to a European, especially German, public—an audience repeatedly shown to lack information about Morocco (Berriane & Popp, 1994). The manager of the Rabat office formulated the strategy on the basis of meetings with the members and consultants of the CG. It was then translated into a call for the competitive submission of designs for the spatial framework of the communication strategy.

The CG's request for proposals by Moroccan architects stipulated that the drafts had to include five areas: "fascination," "information," "environmental protection," "cooperation," and "other aspects." To me, the distinction between fascination and information seemed to have the most significant impact on the subsequent implementation of the stand's strategy. The fascination area was intended to prepare visitors to receive the targeted information and messages they would encounter in the areas that followed. This fascination area "had to evoke the richness of Morocco's cultural heritage, know-how [handicrafts in this case], and history" (*Elle doit restituer la richesse du patrimoine culturel marocain, de son savoir-faire et de son histoire*, Note programme, 1999, p. 2). By contrast, the area dedicated to information was intended to provide a snapshot of the status quo in Morocco so as to fill in assumed gaps in the visitors' knowledge. This section consisted largely of basic information about the country's geographic location, political system, institutions, and its "spirit of tolerance and open-mindedness" (*l'esprit de tolérance et d'ouverture*, p. 2). Given the slogan of Expo 2000, the stand's designers also needed to present Morocco's potential in various realms of life (e.g., tourism, industry, and rural development). In doing so, they were to employ "appropriate communication techniques," such as "audiovisual means" and "modern material of information transmission" (*Il sera fait appel aux outils audiovisuels, au materiel moderne de transmission des informations et de messages*," p. 3).

In other words, the logic behind the design of these two areas of Morocco's stand at Expo 2000 was that the aim of fascinating its visitors would be adequately achieved with culture and history; the aim of informing them, with quantifiable facts and figures.[8] The topoi of national narrative—tradition and modernity—were thereby translated into different spaces. Together, they were intended to project an image of Morocco as a nation, with tradition and modernity as a continuum evolving within Moroccan territory, under its constitutional monarchy, and with optimal prospects and potential for future development.

In turn, the logic behind the technological means for "transporting" these two sets of cultural information was likewise twofold. The content of the fascination area (culture, history, and handicrafts), for which the CG's rules did not specify the exhibition media to be used, obviously suggested a museum-type approach based on objects, commentaries, and practical demonstrations, respectively. In the information area, however, use of modern information technology (IT) was explicitly mandated. That is, the CG's call for design proposals stressed certain affinities between the type of exhibition content and the type of media used to put it across.

The Architectural Execution of the Communication Strategy

The architect's team slightly, but effectively, transformed this clearly defined spatial and media-related execution of the communication strategy by emphasizing the aspects of fascination and animation. The exhibit's resulting representation of Morocco revealed the architect's skepticism that a western perspective would permit recognition of the country's actual (or potential) present. After two inspections of the planned site in Hanover, the architect circulated a revised brochure that proposed two major design changes (Fig. 1).

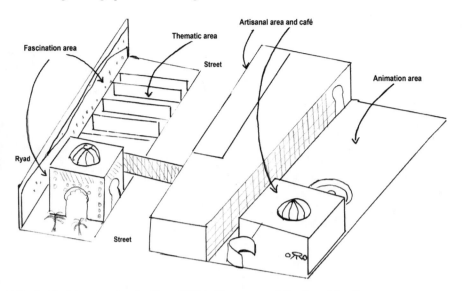

Fig. 1 The Moroccan stand at Expo 2000 in Hanover: spatialization of functions.
Source: author's sketch

Conspicuously, the space-utilization plan made the prescribed information area into a presentation area consisting of four pavilions devoted to the prescribed environmental theme, a stand for the tourist office, and a VIP lounge. Although the stand's most important goal may have been to impart information, the design allotted comparatively little space to that function. The architect explained that the plan was to have the four information pavilions attract the public's attention with state-of-the-art computer animation like that at Disney World. They would depict Moroccan flora and fauna, drawing visitors by surprising them with a novel image of Morocco (see also Expo 2000 Hanovre, 1999, p. 10). This combination of information and technology to create a hyper-real space was not realized, however.

The fascination area, too, was comparatively small in this version of the brochure. The space was to encompass a row of elements making use of symbolic and theatrical effects: (a) a Ryad (an inner courtyard with horseshoe arches and decorated walls enclosing a fountain) reminiscent of city villas; (b) scenery representing an evening sky on the inside surface of the stand's perimeter wall; (c) an intense play of colors on the ceiling, multicolored bolts of fabric, and, again, stage-decor techniques typical of the inner-city weaving district; and (d) a bridge leading to a VIP area on the level above (a space hidden from view by moucharabieh, or projecting second-story windows of latticework). In addition to these eye-catchers, the proposed concept provided for further construction elements of highly symbolic content: the plaza, the bridge, and the street. A row of shops in the style of a traditional handicraft market lined one side of the street. Information pavilions were planned for the other side (Fig. 2).

square and the bridge.

The bridge :

This is the path leading to Morocco, the land of encounter and hospitality. It is the link that ties past to present, tradition to modernity. It connects Africa and the Arab world to Europe.

The square :

This conjures up the Hassan Tour esplanade. The square is the place where every thing converges, a place of constant liveliness, where the visitor, surrounded by handsomely decorated walls, is immersed in an authentically Moroccan atmosphere. Indeed, all the sounds and smells originally encountered in a Moroccan square are faithfully reproduced so that the visitor may find himself experiencing the ever-lasting Moroccan charm.

The street :

This is a passage splitting up the stand into two unequal sets and simulating a medina thoroughfare, hosting various jobs and traditional professions.

The concept likewise harmonises modern volumes and traditional materials. It proves the possibility of combining contemporary architecture and traditional products and symbolises a country where **modernity stems from its very past**. With this end in

Fig. 2 Details and model of the Moroccan stand at Expo 2000 in the press kit.
Source: Hanover Expo 2000 General Management, Morocco section, p. 2

These design elements clearly represented the well-known repertoire typical of the Arab realm as experienced at world's fairs. As constituents of the "fascination" area of Morocco's stand at Expo 2000, they corresponded to what Çelik (1992) called "residential prototypes" (p. 185): "Such a pattern has been common in the architectural representation of Islam in more recent world's fairs that have recycled, for example, the familiar elements that signified North African Islamic architecture to the West for 100 years" (p. 185).[9] Fascination was to be achieved by stage decor taken from urban clichés and combined with illusionary stage decoration whose objective was to create a Moroccan atmosphere. Atmosphere, mood, and symbolic spaces, sometimes borrowing from oriental residential prototypes, were to be an attraction. Architecture would be a vehicle for culture, handicraft, and history.

In addition to this partly implicit use of an oriental archive, the architects pressed for a focus on folklore and animation within this architectural space. They decided to allocate the handicrafts area, the "Moorish café," and "animation" nearly twice as much space as the other areas. The presentation of folklore and handicrafts also relied on prototypes that had emerged and crystallized at nineteenth-century world's fairs. Benedict (1994) distinguishes between three possible practices of exhibiting the ethnic culture of colonized peoples: "the display of people and their artefacts as curiosities, as artisans with their products, and as trophies or booty" (p. 29). Like Çelik (1992), he observed that the formerly colonized states also reproduced the forms of expression and images that world's fairs had always reserved for participants from nonindustrialized countries and colonies: the representation of a cultural (national) unity on the basis of artisans, folklore, and artifacts. Imperialistic ethnocentricity was replaced by national ethnocentricity (Benedict, 1994, p. 55).

Geography of Knowledge—Technologies of Nationhood

Participant observation and detailed examination of the guidelines and plans led me to the assumption that the concentration on animating the stand with folklore and demonstrations of arts and crafts resulted from strategic underestimation of Morocco's potential to portray itself competitively with technological media. The brochure and an interview with the architect indicated the conviction that the Moroccan representation could not match those of the western states on that score and was thus forced to use folklore to create a "warm" image of Moroccan culture. Folklore, unlike high-tech, was understood as a "living" medium:

> In addition to these symbolic elements, Morocco's stand must project a festive image, one of colors and light. It's a convivial atmosphere that visitors are to take with them. Unable to vie with the high-tech pavilions of western countries, it is appropriate to give Morocco a warm atmosphere rather than mount any ice-cold demonstration. (Expo 2000 Hanovre, 1999, p. 11)[10]

Having become acquainted with the future site and with the models of stands being built by other states at Expo 2000, the architect's team working on Morocco's exhibit decided it would be impossible to present an overall image that deviated from this given interpretive framework. The Moroccan participants were convinced of the architect's expert knowledge and by their own impressions of the Hanover building site, so they assumed that one should appeal to the prospective visitor's own visual expectations of Morocco at the world's fair. Avoiding a representation that a western public would count as implausible, the participants ultimately voted for a design that ignored most of what might have been transmitted technologically as knowledge (targeted information and messages) about Morocco's reality as outlined above.

At this juncture, the decision to reproduce the topography of modern world's fairs must be regarded as strategic. Financial concerns about a high-tech representation of Morocco were not the main factor. After all, Morocco had constructed one of the most expensive pavilions at Expo '92 in Seville (see Harvey, 1996, p. 144). More important was the belief that the originally foreseen image would not have been credible. This persuasion derived from a particular geography of knowledge. As an interpretative horizon, the geography of knowledge (based on expert knowledge and experience) represented the spatial distribution of knowledge (understood in this case as the capacity to transmit information by technology) and power.

This geography of knowledge recalls what Harvey (1996) observed at Expo '92, where the frameworks of some national pavilions reflected attempts to break with conventional representation by playing ironically with reflexivity, irony, parody, and prejudices. The resulting self-images had to do particularly with sophisticated exhibition technologies to which the knowledge society at the time promised equal access. Instead, as Harvey asserted,

> the myth of equal access hides or displaces the way in which all have equal access to what are, for some, only images—a range of choices or options which themselves re-inscribe hierarchies of value and reproduce the differences of class and race. For not all can produce image to the same effect, parody is not available to all participants, not all are able convincingly to conflate image with life itself. (p. 126)

Harvey points out that the "powerful" states were playfully referring to their respective national cultures by means of communication technologies, whereas "it was the less powerful nations which were still playing within the earlier frame of reference" (p. 129). This topography harbors a real geopolitical question about power: Is one working with a complex or rather with an uncomplex concept of culture? In other words, is the principle of national culture confirmed or contradicted? Irony, parody, and certain types of images are simply not promising for some countries even if they were to use advanced technology. This negative logic becomes apparent in the most literal sense through the term *technologies of nationhood* (p. 53).

For Morocco's contribution to Expo 2000, technology and perhaps also irony did not constitute adequate strategies for situating the country on the stage of the world's fair, but folklore and arts and crafts did. Imparting information by means of modern technology was indeed something that the planners strove for in the communication strategy. But after reassessment of the situation with the architect on the exhibition site and after consideration of literature on and personal experience with visitors' expectations, informational knowledge was used poorly in the information pavilions, with just single screens showing thematic video-loops (see Fig. 3).

Fig. 3 The Moroccan stand at Expo 2000 in Hanover: thematic pavilion.
Source: author's photograph

One of the factors marking the relationship between technology and Morocco's self-image was the country's own negative assessment of its ability to compete with other states in this area. This self-assessment translated into an enhancement of folklore as a medium, one of whose great merits was a capacity for what the architect called *dépaysement, that is,* for producing an experience that would transport the visitor to another world. The lack of confidence with technological exhibition tools was eventually confirmed by the visitors. They showed little receptiveness to the technologically equipped exhibition areas (which had few screens), but I observed that they had keen interest in those areas animated by live demonstrations of crafts and performances of folklore. Those spaces were among the most popular of the stand. What had been perceived as an asymmetric relationship between north and south was thus translated into a technological equivalent (linking the north with dominance and IT and the south with nondominance and folklore) and was converted into the aesthetic orientalist language that has the greatest affinity for this kind of self-representation (see Fig. 4).

Fig. 4 The Moroccan stand at Expo 2000 in Hanover: view of Ryad and street. Source: author's photograph

An alternative self-portrait had existed, however, for I kept hearing that the members of the jury judging the architectural competition had engaged in controversial discussion over an architectural proposal, dubbed "the kasbah," that was quite the opposite of the design that eventually won out. It was to be a square, dark-colored, plain building entirely closed except for a simple door. In contrast to this unprepossessing exterior, the inside walls consisted of screens displaying video clips in endless loops. As one member of the jury put it,

> There has been a vivid debate about the kasbah. Two or three people thought it was a fantastic idea. The idea was fantastic! But the implementation was too "techno"! It wasn't technical; it was techno—only music, sound, and image. Well, that is surprising. It is very

surprising, but it did not correspond. It was a little sobering, a sober Morocco. (A. Fatmi, personal communication, Casablanca, October 25, 1999)[11]

In this proposal Morocco was nothing but technology, an image considered inappropriate to Morocco and, as shown above, inconsistent with the technologies of nationhood attributed to it within a geography of knowledge.

The Invisibility of "New Smartness"

In this case study I have maintained that assumptions about the performative nature of a given geography of knowledge as developed in the representational work of an interdisciplinary group of professionals in Rabat, Morocco, inscribed that country's exhibit into a classical topography of world's fairs. This example shows that representational work—as a knowledge practice and a performative representational format—draws on assumed geographies of knowledge and thus contributes to their actuality through specific technologies of nationhood.

This reinscription and actualization can be read from two different angles. The first one is the perspective from within the exhibition complex. One recognizes that the Moroccan architect, in designing areas that fascinated visitors with live demonstrations and performances instead of factual and technological representations, reduced complexity in two ways. For one, he reproduced prototypes that effectively plumbed the archive of oriental and world-fair aesthetics, setting Morocco apart through mimicry of a historically overcome colonialist viewpoint. For another, he addressed visitors exclusively as tourists who would deliberately ignore most of Morocco's reality (because they were on vacation). The latter approach momentarily exploited theatrical practice, which one could indeed claim to have helped the Moroccan team project an image of Morocco consistent with the West's own fantasy and then turn Western visitors into paying subjects of it.

I propose to denote the linking of both approaches as a communication strategy that establishes a dramatized form of mimicry as a mode of representation. It appears to me that mimicry and theatrical practice best describe this mode's borders, laws, and especially its scope. Mimicry as an analytical and strategic process based on reflection provides deviation, interpretive variations, and duplications. However, access to this mode of representation is limited. As Bhabha (1994) states, "the excess or slippage produced by the *ambivalence* of mimicry (almost the same, *but not quite*) does not merely 'rupture' the discourse, but becomes transformed into an uncertainty which fixes the colonial subject as a 'partial' presence" (p. 86; see also Taussig, 1993).

Whereas the concept of mimicry emphasizes the aesthetic limits of self-representation, the concept of theatrical practice recalibrates the scope of this strategic representation. Theatrical practice mediates what is presented, clearly showing it not as absolute truth or reality but rather as truth or reality relative to a particular temporal context. This approach proceeds in two ways. For one, theatrical access, being restricted to a particular stock of images, exposes its partiality and dilutes essential meaning attributed to the representation. For another, it temporarily reduces complexity. Both the public and the performers are aware of this specific temporality of the representation. The objectified rituals, divested of their contextual meaning on stage, are perceived and performed as "make-believes" (Benedict, 1994, p. 57). The scope of this representational mode is restricted primarily by the time within which these "make believes" are performed and by the lack of tacit agreement on their fictional character.

The second angle from which to read the reinscription and actualization of assumed geographies of knowledge is the topos of knowledge society. If, as that traditional motif suggests, lack of access to information in the knowledge society is tantamount to ignorance, then the crucial difference between poor and rich becomes the difference between countries that are ignorant and those that are not (see Rötzer, 1999, p. 98, referring to Peter Drucker).

The representational work that I was observing and taking part in was informed by this topos, in which knowledge is understood as a major resource in globally competing societies. The fact that the members of Morocco's group of experts brought in both academic knowledge and knowledge from the realm of civil society as they developed the country's contribution to Expo 2000 indicated two things. First, decentralization of knowledge production is a characteristic of the knowledge society. Second, former centers of modern knowledge production (e.g., the university) face increasing competition from other realms of knowledge production (see Maasen, 1999). If knowledge in the knowledge society has to be practical and oriented to problem-solving, then the subsidiary themes chosen by the group of experts offered appropriate fields for problem-solving knowledge.[12] And, of course, the potential for new forms of exclusion from that kind of global society figured implicitly and explicitly in the actual work and discussion within the office.

It may be no coincidence that the major indicator of a knowledge society was absent from the exhibit itself: the fact that *access* to knowledge, not its possession, was at stake.[13] The acceleration in the development of communication technology in Morocco and the emergence of new and increasingly important forms of knowledge-based labor there were almost invisible in the few videos and leaflets available at the exhibition site.[14] In this sense, demonstrating the cultural competence of being smart, not ignorant, articulates something inscribed in the mechanism of the knowledge society: a form of modernity. According to cultural analyst Andrew Ross (1994), "the new smartness is an advanced form of competition in the sphere of intelligence, where knowledge, more than ever, is a species of power, and technology is its chief field of exercise" (p. 341). The alternative project—the "kasbah" proposal that was rejected in 1999, manifested the three-dimensional representation of such smartness.

But in 2000 the analysis of power relations within the global realm of knowledge that was presented and represented at Expo 2000 made it appear more appropriate to render new smartness invisible than to highlight it. Morocco's former Secretary of State for Post, Technology, Telecommunication, and Information, Nasr Hajji, expressed this kind of smartness perfectly by systematically calling for a leap in history that would allow nations to skip the problems experienced by the former industrial societies and to launch straight into developing the potential of an information and knowledge society: "We might thereby turn our tardiness into trumps because we don't have to solve the problems linked to the sluggishness of industrial society and because we may gain direct access to the most recent technologies (Hajji, 2001, p. 141, author's translation).[15] Murphy's law? In Hanover, Morocco played the world's-fair game by presenting itself as ignorant in the eyes of a public that was inevitably reduced to ignorance, too.

Acknowledgment I am very grateful to the organizers, especially Heike Jöns, and members of the Heidelberg conference on "Geographies of Science" for the inspiring discussion of my paper and to the referees for their helpful comments.

Notes

1. I do not use the term *knowledge society* here as an analytical tool. Instead, I consider the topos "knowledge society" as part of a polarized and performative, discursive field. An overview and discussion of the different positions on the knowledge society's promises of democratization is included in the "Gut zu wissen" collection (Heinrich-Böll-Stiftung, 2002).
2. For a detailed analysis of the contentious local processes involved in preparing Expo '92 in Seville, see Maddox (2004).
3. Avenues for engagement in civil society had been paved mainly since the 1980s (see Leveau & Bennani-Chraïbi, 1996; Monjib, 2003).
4. Though I narrow my argument in this chapter to representational work (the exhibit), the main thrust of my fieldwork was to analyze the construction of knowledge within that context. The anthropology of knowledge (see Knorr-Cetina, 1999; Latour, 1987) led me to ask not only how different disciplinary modes of knowledge were translated into a national representation but also how they would be recognized as knowledge compatible with the aim of representational work. How would they correlate with each other? From the perspective of "knowledge work" (Willke, 1998) within the flexible and competitive format of project work, knowledge has become the main resource and is considered an expression of the knowledge society.
5. As Nadis (2007) comments about the 2005 world's fair in Aichi, Japan, however, these "less dramatic themes of environmentalism and global cooperation" and the "green future" they promise do not have "the appeal of the older offerings. It just isn't as much fun." The mottos for the world's fairs in Aichi ("Wisdom of Nature") and Zaragossa, Spain (2008, "Water and Sustainable Development"), illustrate this restraint, as does the exhibition motto for the world's fair to take place in Shanghai, China, in 2010: "Better City, Better Life."
6. "La participation nationale à l'exposition universelle doit se fonder sur une information fiable et crédible, mesurable et vérifiable dans les faits et sur le terrain, reflétant les atouts du pays et la dynamique en cours pour faire face aux défis du prochaine millénaire."
7. The possible reliability, or rather the attractiveness, of exhibition contents and tools were also tested on me during my fieldwork. I was addressed not only as a Ph.D. student in anthropology doing an internship at the office and writing a dissertation on the subject but also as a prototypical German visitor to Expo 2000 and was asked to test the Moroccan pavilion out at Expo 2000.
8. These facts and figures were expected to translate a certain mentality ("Morocco's open-mindedness," as later stated in the leaflet) into infrastructural terms, such as the number of the country's harbors and airports. Morocco was supposed to become plausible as "a strategic country at the gateway to Europe" (Generalkommissariat, 2000a, p. 3).

9. By 1930, a rather generic Oriental-Islamic world-fair architecture had emerged—structures in the form of mosques, cafés, bazaars, administrative buildings, obelisks, and temples. Regarded as attractions of every world's fair, these architectural prototypes were repeatedly built and continued to be effective until Expo '92 in Seville. The Moroccan pavilion for world exhibitions followed in the same tradition. It was notable for its exotic and fitting exterior modeled on mosque designs and for an interior characterized by an exhibition architecture in modern museum style.

10. "Outre ces éléments symboliques, le stand du Maroc doit imprimer une image de fête, de couleurs et de lumière. C'est une ambiance conviviale que le promeneur doit garder avec lui. Ne pouvant lutter avec les pavillons à haute technologie des pays occidentaux, il convient de donner au Maroc une atmosphère chaleureuse à l'opposé de tout démonstration froide glacée."

11. The name of the interviewee has been changed.

12. In Rabat, each member of the Moroccan Expo 2000 exhibition team made an effort to gain recognition of his or her expertise. Knowledge as a resource therefore resided in highly analytical project-based, representational work, flexible management of one's own career, the availability of information, and the transformation of information into knowledge within the code of applicability. For a detailed analysis of ways to identify knowledge as a resource in this context, see Färber (2006, pp. 69–148).

13. This circumstance was apparent, for instance, in the competition for access to information and the demonstration of this access in the daily work at the office.

14. See, for example, the brochure *Marocco: The Roots of the Future* (Generalkommissariat, 2000b) for a statement about the country's infrastructure: "Morocco, more than ever before, must assume its role as an international link. Accessible from all over the world thanks to its twelve international airports and its large Mediterranean and Atlantic ports, it will soon be directly linked to the European rail and road networks by means of a permanent connection, a tunnel, linking the city of Tangiers with the south of Spain" (p. 14).

15. "Nous pourrons alors transformer nos retards en atouts, puisque nous n'avons pas à résoudre les problèmes liés aux pesanteurs de la société industrielle, et puisque nous pouvons accéder directement aux technologies les plus récentes."

References

Avant projet [Before the project]. (1999, August). (Unpublished internal documents).

Benedict, B. (1983). *The anthropology of world's fairs: San Francisco's Panama Pacific International Exhibition of 1915*. London: Scolar Press.

Benedict, B. (1994). Rituals of Representation: Ethnic Stereotypes and Colonized Peoples at World's Fairs. In R. Kroes (Series Ed.) & R. W. Rydell & N. E. Gwinn (Vol. Eds.), *European contributions to American studies: Vol. 27. Fair representations: World's fair and the modern world* (pp. 28–61). Amsterdam: VU University Press.

Bennett, T. (1988). The exhibitionary complex. *New Foundations, 4,* 73–102.

Berriane, M., & Popp, H. (Eds.). (1994). *Die Sicht des Anderen. Das Marokkobild der Deutschen, das Deutschlandbild der Marokkaner* [The view of the other: How Germans see Morocco, how Moroccans see ermany].Passau, Germany: Passavia Universitätverlag.

Bhabha, H. K. (1994). *The location of culture*. London: Routledge.

Çelik, Z. (1992). *Displaying the orient: Architecture of Islam at nineteenth-century world's fairs*. Berkeley: University of California Press.

Expo 2000 Hanovre. Stand du Maroc [Expo 2000 Hanover: Morocco's exhibit]. (1999, August). (Unpublished architect's proposal).

Färber, A. (2006). *Weltausstellung als Wissensmodus. Ethnographie einer Repräsentationsarbeit* [The world exhibition as a mode of knowledge: An ethnography of representational work]. Berlin: LIT Verlag.

Generalkommissariat. (2000a). *Driving forces for development.* Hanover: Generalkommissariat (Press Brochure).

Generalkommisariat. (2000b). *Marocco: The roots of the future.* Hanover: Generalkommisariat (Press Brochure).

Gilbert, J. (1994). World's fairs as historical events. In R. Kroes (Series Ed.) & R. W. Rydell & N. E. Gwinn (Vol. Eds.), *European contributions to American studies: Vol. 27. Fair representations: World's fair and the modern world* (pp. 13–27). Amsterdam: VU University Press.

Hajji, N. (2001). *L'Insertion du Maroc dans la société de l'information et du savoir. Pour une nouvelle vision* [The insertion of Morocco into the information and knowledge society]. Casablanca: Editions Afrique Orient.

Hammoudi, A. (1997). *Master and disciple: The cultural foundations of Moroccan authoritarianism.* Chicago: University of Chicago Press.

Harris, N. (1990). *Cultural excursions: Making appetites and cultural tastes in modern America.* Chicago: University of Chicago Press.

Harvey, P. (1996). *Hybrids of modernity: Anthropology, the nation state and the Universal Exhibition.* London: Routledge.

Heinrich-Böll-Stiftung (Ed.). (2002). *Gut zu wissen. Links zur Wissensgesellschaft* [Good to know: Links to the knowledge society]. Münster, Germany: Waxmann Verlag.

Khatibi, A. (1998). *L'alternance et les partis politiques* [Broad change and political parties]. Casablanca: Eddif.

Knorr-Cetina, K. (1999). *Epistemic cultures. How the sciences make knowledge.* Cambridge, London: Harvard University Press.

Latour, B. (1987). *Science in action: How to follow scientists and engineers through society.* Cambridge, MA: Harvard University Press.

Leveau, R., & Bennani-Chraïbi, M. (1996). Maroc 1996: Institutions—Economie—Société [Morocco 1996: Institutions—Economy—Society]. In H. Bozarslan, R. Leveau, & M. Bennani-Chraïbi (Eds.), *Acteurs et espaces politiques au Maroc et en Turquie* (pp. 47–104). Les Travaux du Centre Marc Bloch 8. Berlin: Centre Marc Bloch.

Maasen, S. (1999). *Wissenssoziologie* [Sociology of knowledge]. Bielefeld, Germany: Transcript Verlag.

Maddox, R. (2004). *The best of all possible islands: Seville's Universal Exposition, the New Spain, and the New Europe.* New York: State University of New York Press.

Mitchell, T. (1989). The world of exhibition. *Comparative Studies in Society and History, 31,* 217–236.

Monjib, M. (2003). Le Maroc: Monarchie, opposition, elites et reforme. Essai d'histoire immédiate [Morocco: Monarchy, opposition, elites, and reform—An essay on immediate history]. *Revue d'Histoire Maghrébine, 30,* 535–542.

Nadis, F. (2007). *Nature at Aichi World's Expo 2005.* Retrieved 08. 10. 2009 http://www.historyoftechnology.org/eTC/v48no3/nadis.html

Note de synthèse. (1999, August). (Unpublished manuscript of the Generalkommissariat).

Note programme. (1999). (Unpublished manuscript of the Generalkommissariat).

Press, O. (2000a, May 31). *Opening address by German Chancellor Gerhard Schröder.* Retrieved August 11, 2009, from http://www.expo2000.de/dp/gesellschafter/rede.html

Press, O. (2000b) *Rede von Bundespräsident Wolfgang Thierse zum Abschluss der Expo 2000, 31.10.2000 [Closing address by Wolfgang Thierse, President of the German Bundestag, October 31, 2000].* Retrieved August 5, 2009, from http://webarchiv.bundestag.de/archive/2005/0718/bic/presse/2000/pz_001103.html

Ross, A. (1994). The new smartenss. In G. Bender & T. Druckery (Eds.), *Culture on the brink: Ideologies of technology* (pp. 329–341). Seattle, WA: Bay Press.

Rötzer, F. (1999). *Visionen für das 21. Jahrhundert: Band 6. Megamaschine Wissen. Vision: Überleben im Netz* [Visions for the twenty-first century: vol. 6. Megamachine knowledge: The vision of surviving on the Net]. Frankfurt am Main: Campus.

Rydell, R. W. (1984). *All the world's a fair. Visions of empire at American international expositions, 1876–1916*. Chicago: University of Chicago Press.

Rydell, R. W. (2006). World fairs and museums. In S. Macdonald (Ed.), *A companion to museum studies* (pp. 135–151). Malden, MA; & Oxford, England: Blackwell.

Taussig, M. (1993). *Mimesis and alterity: A particular history of the senses*. New York: Routledge.

Willke, H. (1998). Organisierte Wissensarbeit. *Zeitschrift für Soziologie, 27*, 161–177.

Part IV
Science and the Public

Geographies of Science and Public Understanding? Exploring the Reception of the British Association for the Advancement of Science in Britain and in Ireland, c.1845–1939

Charles W. J. Withers

Much recent work on the nature, making, and reception of scientific knowledge and its variant disciplines—including geography—has drawn attention to the importance of the spatial setting (for summaries, see Finnegan, 2008; Livingstone, 1995, 2003; Naylor, 2005; Powell, 2007; Shapin, 1998; Smith & Agar, 1998; Withers, 2001, pp. 1–28; Withers, 2002). Some of this research investigates the diverse sites of science's production, such as the ship (Sorrenson, 1996), the botanic garden (Spary, 2000), or the laboratory (Kohler, 2002). Some of it concentrates on the sites of science's reception, including the different social spaces of scientific reading and translation (Rupke, 1999; Secord, 2000). Still other studies tackle questions to do with the mobility of science (Secord, 2004) and the performance of science, including its oratorical cultures and speech sites (Livingstone, 2007; Secord, 2007). Regardless of the focus, though, there can be no doubting the central place now occupied by the "geographical turn" in understanding the ways in which science has been made, circulated, and received.

This chapter draws upon such inquiry and is intended as a contribution to it. Its focus in general is the historical geographies of the British Association for the Advancement of Science (BAAS). Of particular concern are questions pertaining to the reception of BAAS annual scientific meetings given that, from its foundation in 1831 and as an aim, the Association was peripatetic, moving to different towns and cities throughout Britain and Ireland in order to promote science as a form of civic good. Because the Association moved annually and because its science consisted of different sections, such as mathematics, chemistry, geology, and geography, there is at least the potential to investigate the ways in which different urban audiences received BAAS science and, perhaps, the differences between subjects in different towns and cities.[1]

Questions about the reception of science are, however, difficult to answer, perhaps even to pose. Audience studies tend to focus either on numbers alone, neglecting the cognitive dimensions of comprehension, or on individual instances,

C.W.J. Withers (✉)
Institute of Geography, University of Edinburgh, Edinburgh EH8 9XP, UK
e-mail: c.w.j.withers@ed.ac.uk

P. Meusburger et al. (eds.), *Geographies of Science*, Knowledge and Space 3,
DOI 10.1007/978-90-481-8611-2_10, © Springer Science+Business Media B.V. 2010

making generalization problematic, not least given the social difference among recipients and the frequent modification of science precisely to meet the needs of such different audiences (e.g., science books for children). Or they have been concerned with readership and reviewing as interrogative forms of reception and so are based largely on the hermeneutics of textual evidence (see, for example, Cooter & Pumfrey, 1994; Keighren, 2006; Rupke, 1999; Secord, 2000). Further, conceptualizations of the public understanding of science in historical context have tended to address the problem of science's reception and comprehension in terms of a deficit-democracy model (Shapin, 1990). In that model, the public—itself a problematic word—is without scientific knowledge (in deficit) or is wholly informed (as part of a democracy) even when it may not understand the details of the science in question. Such a "canonical account" of the public understanding of science, as Shapin noted, needs to be scrutinized in different geographical and historical contexts. As other research on the public's engagement with science shows, audiences vary in their knowledge. In any case they have tacit understanding and expertise and vary in their response not just to the science but to its political implications and to the scientist and scientific institutions as credible people and places to be trusted or not (Collins & Evans, 2002; Gregory & Miller, 1998; Yearley, 2005). This chapter does not deal with all of these issues. But, in looking at the ways in which BAAS science was received, I want to suggest that it is possible to deepen understanding of the workings and effects of a scientific organization that, in its nineteenth-century heyday, saw itself as the "Parliament of Science" (MacLeod & Collins, 1981, p. 1). I submit that this approach can contribute to studies in the historical geographies of science's reception.

Studying the Reception of the BAAS: Toward Urban Historical Geographies of Provincial Science

The BAAS was begun in York in 1831 and continues today, chiefly through its annual meetings, as an organization geared to the public dissemination of scientific knowledge. The objectives of the BAAS at its foundation were—

> [t]o give a stronger impulse and a more systematic direction to scientific enquiry; to promote the intercourse of those who cultivate Science in different parts of the British Empire with one another and with foreign philosophers; to obtain more general attention for the objects of Science and the removal of any disadvantages of a public kind which impede its progress. (Opening paragraph of the BAAS Constitution, as cited in MacLeod & Collins, 1981, p. i)

Earlier work on the BAAS has considered its initial years (from 1831 to about 1845), its leading individual members, and its institutional mechanics or has inquired into the connections between BAAS science and imperialism in the late nineteenth and early twentieth centuries (e.g., Morrell & Thackray, 1981, 1984).

Historians of the Association have revealed close affiliations between its members and the civic promotion of science in provincial Britain, especially for the organization's first decade and a half. They have suggested, too, that the BAAS's democratic and meritocratic mission was to a degree compromised, at least before the early 1850s, by the dominance of aristocratic figures and other "gentlemen" of science. Those within the Association saw it as an important means to promote science as a form of cultural good and regarded the Parliament of Science as an important civic institution (MacLeod & Collins, 1981; Morrell & Thackray, 1981, 1984; Worboys, 1981).

More recent studies of the BAAS have examined the work of its Section E, geography, in the century from 1831 on and explored the several vehicles through which the BAAS acted to promote science after 1845. These were its annual meetings, regional and local handbooks, lecture programs, excursions, and financial support of particular research themes, which often brought together more than one BAAS section (Withers, Finnegan, & Higgitt, 2006; Withers, Higgitt, & Finnegan, 2008). Such work has begun to highlight the particularities of the different ways in which science was made in given urban settings. It has thereby contributed to a growing body of critical analysis of the historical geography of urban science (Dierig, Lachmund, & Mendelsohn, 2003; Gieryn, 2006) and, together with that literature, has signaled how much more there is still to know about the nature and reception of science in historical context.

The sources available to document the reception of BAAS scientific meetings in different urban settings vary in type and in their survival. Attendance figures were seldom given and never consistently kept. Diaries permit insight into individuals' engagement with certain meetings and topics—and are drawn upon in what follows—but generalization is always difficult from individual accounts of this kind. What the BAAS archive does contain, however, is an extensive newspaper record, mainly of reports on each of its meetings and on the place of the BAAS in the town in question. Such was the interest in visits of the BAAS and its program of activities that some towns went so far as to collate different newspaper reviews of BAAS science as a guide to its annual meetings (as, most notably, in Bath in 1864). Most towns had their meetings covered by local newspapers and the major national papers. Of course, the use of newspaper accounts warrants caution because they in effect offer an interpretive mediation, via the reporter or editor, between the events themselves and the views of the recipient audience, the precise degree (and purpose) of which may not be recoverable. But in combination with other sources, particularly when different papers make similar claims concerning the reception of science, newspaper coverage can illuminate how the meetings were received, which sectional subjects were the most popular, and, occasionally, why and even how different groups within the local audiences for BAAS science reacted. Drawing upon newspaper evidence and other printed material as suggestive rather than definitive illustrations, I explore four interrelated themes in the reception of the BAAS: professional reaction, scientific differences, civic and political context, and gender differences.

Professional Reaction

Because the BAAS operated via different sections (Section A: Mathematics and Physics, B: Chemistry, C: Geology, D: Zoology, E: Geography, F: Economics, G: Engineering, H: Anthropology, I: Physiology, J: Psychology, K: Botany, L: Education, and M: Agriculture), it ought to be possible to trace the reception of BAAS meetings, particularly of subject-specific sectional matters, in those subjects' professional journals. Let me take geography as an example.

Reports on BAAS meetings and Section E proceedings regularly appeared in geography journals, notably the *Geographical Journal* (the outlet of the Royal Geographical Society) and the *Scottish Geographical Magazine* (the outlet of the Royal Scottish Geographical Society). In most instances, this type of coverage consisted of brief summaries of the president's speech and the content of papers delivered in the sectional program. On topics of continuing concern, however, such as African exploration, polar work, survey work, and imperialism, or of interest deriving from connections between the town in question and the content of the lecture program, societies' officers often contributed additional comment on the reaction to presented papers. Further, there were at times responses to the presentation of geographical material within the programs of other sections. Reporting upon the 1887 Manchester meeting, for example, Mill (1887) praised Sir Francis de Winton's evening lecture on African exploration as "a model of what a popular lecture should be. He handled his familiar theme with a systematic clearness and quiet enthusiasm which secured the sympathy and unfaltering interest of his large audience" (p. 525). Mill also observed how the meeting "was of more than usual interest to geographers, for, besides a number of valuable papers read to section E, there were many bearing directly upon the subject submitted to other sections" (p. 526). Noting, too, how "[v]ery important work was got through on Monday" (p. 527)— including discussions on the scope and teaching of geography and on the work of the Ordnance Survey—Mill documents scenes of extremely animated debate, hurried papers because of lax timetabling, and, *en passant*, a poor-quality BAAS meeting handbook, "a sorry contrast both in bulk and quality to the Birmingham hand-book of last year" (p. 529).

In the case of geography as a BAAS science, one consistent theme in its reception was the difference *within* that subject between the technical papers, normally given by professionals and experts, and the popular papers, usually treating a topic of exploration. The following report upon the 1894 Oxford meeting stands as one of many examples of this point—and of much else—about the reception of geographical science:

> The attendance was from first to last exceptionally good, and the audiences held together with quite remarkable constancy. That, however, was evidently due to the popular rather than to the strictly scientific character of the papers and discussions. It must always be remembered that in the British Association two audiences have to be catered for—the scientists and the amateurs—and that the financial success of the meeting depends much more on the latter than on the former. Complaints are often made—they have been made this year—that the treatment of the subjects in some sections is so technical that it repels the lay

members, who form the bulk of the membership of the association. The objection may or may not be deserving of notice; but it certainly does not apply to the Geographical Section. That section is the happy hunting-ground of the unattached and amateur Associate. Thanks to the profuse and promiscuous use of the magic lantern, it has become the attractive show-room of the Association. But geography as a science, or the scientific aspect of geography, does not gain much by this ephemeral popularity. The audience is panting for sensations; the ubiquitous and irrepressible globe-trotter is the ideal of the hour, and the sensation is all the greater if the globe-trotter happens to be a woman. The paper which attracts a crowded audience may be a tedious narrative of a holiday spent in Armenia, or in Mexico, or in the desert of Libya, or in Montenegro, or in Arabia; but all its sins are forgiven if it is illus-trated by what the official programme calls "optical projections", which means, in common parlance, "lantern slides." (Dalgleish, 1894, p. 463)

Reports in professional journals must, of course, be used circumspectly in assess-ing science's reception. The personal interests of the reporter and the purpose of the report need to be acknowledged. In a sense, too, such reports are a form of sci-ence's production, or perhaps reproduction, rather than its reception alone because, for persons unable to attend, they provide a record of what took place and how it was received. Professional etiquette and disciplinary conventions could mean that journal reports did not disclose the contested nature of science's making and recep-tion or did not ever capture the nuances of the spoken delivery, the reactions of the audiences, and so on. Reporting upon the 1911 Portsmouth BAAS meeting, for example, the anonymous commentator in the *Scottish Geographical Magazine* noted only that the president's address to Section E (by Colonel Charles Close) "did not meet with entire acceptance by the geographers present" (Meeting, 1911, p. 517). The full reaction to Close's views—simply, that geography was not a sci-ence yet needed to be one—was never made public. The *Geographical Journal* did not even report the speech, preferring to concentrate on the technical papers. Other geographers sought to keep their disagreement from becoming public. Indeed, they persuaded the reporter from *The Times* (London) to omit this element in his cov-erage of the meeting—even though, as private correspondence reveals, many were greatly agitated and held different views on how to respond to Close's controversial remarks (Withers et al., 2006, pp. 444–446).

Scientific Differences

In planning its meetings in negotiation with local bodies and prominent individuals in given towns, the BAAS used local or regional particularities to give structure and specific content to its excursions or elements of its sectional programs. It made use of the coal and textiles around Manchester in 1868, for example, and of the coal and iron-working in South Wales as an attraction for the metallurgists in Section B at the Swansea meeting of 1848 (Miskell, 2003). For the members of Section G, the BAAS drew on Glascow's many engineering works when it visited that city in 1876 and 1901. In port towns, such as Liverpool (1923), Hull (1922), and Newcastle (1889), the organizers of the Section E program set out to incorporate papers on geography's commercial importance given these cities' place in Britain's imperial trading networks. The BAAS may, then, have been peripatetic, but its mobility neither produced nor depended upon a simple and single form of "BAAS science." Local circumstances mattered in shaping what was given in the proceedings.

Nevertheless, there is consistent evidence that certain subjects were better attended and better received than others and, as with the above example of geography in the Oxford meeting of 1894, that differences existed within individual sciences between the more popular papers and the more technical or "expert" ones. Geography was consistently well attended, notably when a famous explorer would present accounts of his travels, as was the case with David N. Livingstone and the Zambesi expedition at the meetings in Bath in 1864 and Brighton in 1872. The popularity of geography as a lecture subject also had a more general explanation. With that field's popular papers, as one Sheffield reporter once put it, "little or no antecedent scientific knowledge is necessary to enable the listeners to comprehend all the points in the memoirs read" (*Sheffield Daily Telegraph,* August 22, 1879, p. 4). In contrast, attendance in the mathematics and physics section went beyond "sparse" only if experimental apparatus was employed. Geology "was of little interest to those who have not a technical knowledge of the subject"; economics was only "moderately popular." Attendance in the anthropology section varied in relation to the topics, being good for "various learned men discussing 'the ape-origin of man'" (i.e., Darwin's theories of natural selection and evolution) and otherwise poor. In Dublin the previous year, Huxley had drawn huge crowds, but not many stayed to hear others' talks: "God save King Huxley rang from every throat, and the hero worshippers trooped apace towards the footstool of the great prophet of biology in the medical buildings. But after him the deluge, or rather the evaporation" (*The Irish Times,* August 17, 1878, p. 3).

Given the concurrent nature of sections' programs, BAAS visitors commonly migrated from one subject to another if a subject or speaker failed to hold their attention. Consuming science effectively demanded a sort of social and intellectual "circuitry," for people moved from topic to topic, room to room, scientific display to *conversazione.* One member of the BAAS audience in Dublin in 1908 explained the encounters succinctly: With the matter of errant scales on maps of the British Isles "not seeming a matter of urgent importance, we drifted into the geological section" (*The Irish Times,* September 4, 1908, p. 12). Indeed, attendance at BAAS meetings was, for many people, not about science at all. Instead, it was often about participating in a civic social gathering and being seen by one's peers to be the sort of person who *ought* to take an interest in science, even if much of it, particularly in certain sections, was never engaged with or, if engaged with, not well understood. Such claims are apparent in the words of the reporter on the Dublin meeting of 1878. Describing what he called "perambulatory science"—his movement across the meeting as a whole—he conveyed the experience of differential comprehension, social bustle, and the limited oratorical capacities of BAAS speakers in their different speech sites:

> Some orators were holding forth, under discouraging circumstances... [O]nly a few [of the audience] seemed to pay any attention, and many had on them a half-amused, half bewildered look of astonishment at finding themselves in such society at all. A geologist

spoke about a mountain range, and revealed the marvels of its hidden mysteries in weird
sentences, and young and old ladies and gentlemen lounged in and out again as ignorant as
before. (*The Irish Times*, August 19, 1878, p. 7)

The reporter depicted the evening soirée quite differently in this article: "[H]ere
science was popularised and beautified, everything brilliant and gay went side by
side with the most marvellous inventions and the deepest research; and harmony,
elegance, and grace added their charms to the brightness and animation of the
scene." His view was no doubt colored by the fact that, as he put it, "We were of
those who got seats, and not amongst the struggling multitude who surged around
the portals of Professor Huxley's room" (p. 7).

Civic and Political Context

The reception of the BAAS could depend upon an audience's perceptions of the
nature of the Association and, from that, upon different civic views as to the purpose of science. In Glasgow in 1876, for instance, leading university figures and
men of science in the town thought the meeting a great success. By contrast, many
townsfolk reckoned it "exceptionally dull" and, therefore, a failure. It ought to be
remembered, noted one reporter, "that the object of the Association is not to amuse
the public, but to advance science, and that it may happen that the most exciting meetings may be the least productive of good results" (*The Glasgow News*,
September 14, 1876, p. 5). The article continued: "There was no morning, for example, so popular as in that in which we were told about the dancing tables, and the
spirits that animated gin-and-water, and in which Wallace and Carpenter [a reference
to the debates between Alfred Russel Wallace and the physician Alfred Carpenter
over ethnology and Darwinian thinking] battled with words and still more with eyes,
yet perhaps there was no morning more conspicuous for its absence of science"
(p. 5). To the editor of the Edinburgh-based *The Scotsman*, there was little point in
having the BAAS come to one's city if there was to be "nothing in the nature of a
quid pro quo meant or involved" (September 14, 1876, p. 6). Although "[t]he inhabitants of Glasgow, as might have been expected, had fed them well, and in every way
treated them as handsomely as they could," it was nevertheless the case that "[t]o the
great majority of the outside public, the whole proceedings, with the exception of
the festivities, must have seemed a sort of intellectual jungle, into which to plunge
is to be lost" (p. 6).

The sense that BAAS science (or perhaps more accurately, BAAS scientists)
could be welcomed and deemed worth going to see and hear if it (or they) afforded
amusement more than edification stemmed partly from the expectation that the local
citizenry provide the board and lodging for BAAS visitors. They would be issued
with lists of hotels and bed-and-breakfast establishments and of places in the city
open to them. Neither the extra influx nor, sometimes, the timing of the BAAS
meeting was always welcomed by city authorities even when they had been actively
promoting their city as an appropriate venue for civic science. BAAS officials

thought their organization slighted in 1871 in Edinburgh, for example, because too few residents offered accommodation. Edinburgh Council members responded that the BAAS had been advised about timing: "The authorities of the British association were *over and over again* warned that if the association visited Edinburgh during either August or September its members would be sure to find the city deserted by those who, had they come earlier or later, would willingly have accommodated them" (*The Daily Review,* August 9, 1871, p. 5).

In Ireland, the BAAS was received with animosity by some not because of the science but because it was British. The meeting and its week-long events of urbane sociability were associated with political ascendancy. Such enmity consistently pervaded the reporting on reception of the Dublin meetings. In 1835, for example, at a time of unrest given protest in southeast Ireland against absentee landlordism and a rising sentiment against Protestant rule, the local organizers of the BAAS meeting banned certain sections of the press from attending and reporting on it. This proscription did not prevent the Catholic and nationalist *Freeman's Journal and Daily Commercial Advertiser* from recording how "the managers of the affair in Dublin, who are a clique of bitter Orangeists," manipulated the reception of the meeting by ensuring that coverage was carried only in Protestant papers (August 11, 1835, p. 11). "[T]he Orange newspapers, which are patronised by the committee of mismanagement, will give ample details, as they have hitherto been alone favoured with the advertisements" (p. 11). The "miserable, consumptive Orange clique" was in fact the College of Physicians. By contrast, the College of Surgeons showed "no niggardliness, no paltriness, no exclusiveness" in hosting a scientific breakfast for leading Irishmen of science, whatever their religious and political leanings.

Reporting upon the 1857 meeting, the strongly pronationalist newspaper *The Nation* urged visitors to recognize the standing of Dublin's scientists and to understand the broader connections between science and national identity.

> Among the savans [*sic*] now assembled in the metropolis of Ireland, there are many whose names are emblazoned on the golden roll of Philosophy, Science and Art. Such men ought to be welcome in Ireland: here they tread a land which was once the home of learning, the munificent patroness of the Arts and Sciences, ere the country for which the Association takes its name, had emerged from the night of barbaric ignorance. Here they will finds traces of all that interest the ethnologist, and the antiquarian; they will find relics of a glorious past, evidence of a miserable present. They may employ themselves profitably in investigating the cause of this state of things; in tracing the date at which this decadence set in, and they will find that when Ireland ceased to be independent, the Arts and Sciences fled the land. (*The Nation,* August 29, 1857, p. 17)

At the 1908 Dublin meeting, the advertisement that hung outside Trinity College was defaced—"maliciously damaged to read 'THE BRITISH ASS',," as the Dublin-based *Irish Times* had it (September 3, 1908, p. 2). Within the meeting, however, the fraternity of science was perceived to overcome factional differences: "It is not the smallest of our debts to the British Association that it provided us with a platform on which Irishmen of all parties have been able to exchange views in the true spirit of scientific enquiry" (*Irish Times,* September 10, 1908, p. 3).

In various ways, then, what Livingstone (1995) has termed "the political topography of scientific commitment" to science's making in urban spaces that follow (or do not follow) "the contours of political persuasion" (p. 19) had its parallels in what may be thought of as the civic topography of scientific reception. Its national and local dimensions, institutional expression, and intrascientific differences demand attention if one is to document how people engaged with science's performance and if it is important to know what it meant for the urban audiences, visitors, and perambulatory participants.

Gender Differences

If the reception of BAAS science varied by place, topic, the distinction between "expert" and "popular," and even presentation (the question of whether lantern slides and other demonstrative apparatus were used), it is also the case that it varied strongly by gender. This is not to say that the observed difference between expert and popular talks was simply one between professional men and unknowing women. After all, the BAAS meetings between 1867 and 1911 included lectures directed at working men, and many of those presentations attracted large crowds. However, the place of women within the BAAS varied over time and in different urban locations, particularly with regard not only to the audience and, therefore, the reception of science but also to women's role as participants and, albeit in limited numbers, as producers of science (Higgitt & Withers, 2008).

At the foundation of the BAAS in 1831, women were formally excluded from the Association's paper sessions but were encouraged to accompany their scientific husbands and relations in order to add social status and glamour to the meeting. Women did attend sectional meetings, though, and, as of 1838, they were formally admitted to all sections except Section D, which, in addition to botany, then included zoology, a subject deemed unsuitable because of its attention to matters of biological reproduction. By 1839, however, women could attend all sectional sessions but only in certain parts of the sites for science—"the galleries only or railed-off spaces" (Morrell & Thackray, 1984, p. 301)—where that segregation was possible and enforceable in the civic buildings in question.

Morrell and Thackray (1981) state that the presence of women was essential in the early years of the BAAS meetings, asserting that they made a vital financial contribution and gave the meetings social status. Research on women's interaction with the BAAS in the period after 1845 confirms these points and suggests that women came almost to define BAAS audiences in certain towns and cities and with respect to the activities of certain sections (Higgitt & Withers, 2008). The reception of geography, anthropology, and, on occasion, geology was colored by reporters' perceptions that women in particular went to see the speaker, not to hear the science: Putative scientific reception was in fact fascination with celebrity.

As others have shown, however, science as a form of cultural good in the ninenteenth century was in a competitive marketplace as perhaps never before (Fyfe &

Lightman, 2007b; Lightman, 2007). "Popular" science was often more popular than scientific. With illuminated lectures and managed displays, audiences could be dazzled by the showmanship more than informed by the principles (Gooday, 2007), or, especially with women, they could be made to engage with only static displays of sciences such as botany (Shteir, 2007). Assumed differences of this kind and the fact that women did indeed occasionally flock to BAAS meetings and lectures to hear such figures as John Herschel, Michael Faraday, David N. Livingstone, and Thomas Huxley led to women's frequent portrayal as being either especially interested in or bored by certain subjects. The effect was that organizers constructed the program to cater to these perceptions. In that sense, preconceived notions of reception did guide what was produced at BAAS meetings. But sectional differences complicate this picture. Botany, geology, and notably geography were considered appropriate female topics. Section A was reckoned hard and masculine. Section F and, from 1901 on, Section L (education) attracted a large number of women as audience members and, proportionate to other sections, as speakers. But the inclusion of women in those two sections, and the topics of social and welfare reform addressed therein, led many people to see such topics as not suitable for the BAAS. In some cases, diary evidence bears out these perceptions. Finding zoology "so uninteresting" and being "rather tired of Geology," Caroline Howard, for instance, headed to see Livingstone in Dublin in 1857, only to be disappointed that he spoke so softly, having strained his voice giving sermons in the African outdoors (National Library of Ireland, MS 4792, Wicklow Papers, Journal of Caroline Howard, 28 August 1857 [no pagination]). For Helen Shipton in Edinburgh in 1892, the science was unimportant. She attended only with the purpose of "Looking on!" or "studying character—a study for which there is here a wide and interesting field" (Shipton, 1892, p. 84).

Notes Toward a Conclusion

Fuller discussion of the issues raised in this chapter should include more detailed scrutiny of the terms used, including *reception, production, audience, public,* and, of course, *science,* and of the relationships between them. Further, it is important to consider the thesis that receptive connections between science and the public must always be read in relation to historical contingencies (Shapin, 1990)—contingencies of period, patronage, and temporal changes in the channels of communication between science and the public. If one accepts it, then I suggest that it is just as important to understand those connections as geographically contingent. That is, it is important to understand them as shaped by spatial differences, by the nature of the setting in which science was conducted and received, and by differences in who took part in science where, not as a producer but as an audience member. Reception and public understanding are closely related matters, but they are not the same thing. This claim echoes others' views about the ways that questions of scientific reception bring into starker relief not only the places in which science is made for the public but also the practices of display, orality, and other performance through which nineteenth-century science marketed itself (Fyfe & Lightman, 2007a).

Against the backdrop of these bigger issues, this examination of the reception of the BAAS meetings has revealed several features of interest. Public understanding of BAAS varied depending on the subject, gender, and (less readily captured in detail in this chapter) the general categories of "expert" and "lay" or "popular" audiences. There is evidence to suggest, however, that it was the performance of science—its associated display, the oratorical power of the speaker—and less its content that drew people to BAAS science, or, more precisely, drew particular people on a particular day to a particular place for science.

Sites for science were also experiences about science: "Not only may a site be experienced differently by different groups of visitors at any one point in time, but visitors at different points in time are almost certain to have different experiences" (Fyfe & Lightman, 2007b, p. 5). In several instances within the BAAS meetings, however, the status of the speaker was more crucial in attracting an audience, at least for certain subjects, than was the content of the talk. Comprehension was deemed less significant than attendance, for women especially so.

Furthermore, there were different civic perspectives on the "value" of BAAS science as a form of local cultural capital and as a function of the week-long presence of the Association in any one year and the fact that it might not return for some years. As one reporter noted:

> The association may, in fact, be compared to a gigantic boa-constrictor, which takes one huge meal a-year, and lies in a semi-dormant state during the rest of the period.... The energy is magnificent, but, at the same time, discontinuous and spasmodic, and though the inhabitants of cities such as Dundee or Bradford may for once in a generation receive a visit from the association, for twenty or twenty-five years they are left to grow up—and they do grow up—in ignorance of the very existence of this great peripatetic body. (*The Scotsman*, August 9, 1871, p. 1).

The result of this brief engagement with questions of reception and public understanding is that the notion of the BAAS as a single institution committed to disseminating so simple a thing as "science" to so uniform a thing as "the public" fractures when the reception of its work is scrutinized. The fact that such questions are a theme central to much recent research in the history of science suggests that the BAAS was, as its founders hoped, an institution key to the making of science and important in giving "general attention" to the objects of science. It also indicates that, for the contemporary researcher, there is merit and more work yet to do in looking at the ways in which BAAS science was embraced by its publics in their several settings. Perhaps, too, other scientific institutions in other times and other places could be considered in these ways for what they might reveal about the complexities of the historical geographies of science.

Acknowledgment I am grateful to the editors for the invitation to contribute an essay to this volume and to the referee for thoughtful comments on an earlier draft, to the Bodleian Library and the British Association for the Advancement of Science for permission to quote from documents in their care, and to the staff of numerous archives and university libraries holding BAAS material for their assistance. This paper is the result of an ESRC-funded research grant entitled "Geography and the British Association for the Advancement of Science, 1831–1933" (ESRC Ref RES-000-23-0927), and it is a pleasure to acknowledge this support.

Note

1. The reception of BAAS science in its seven overseas meetings (Montreal in 1884, Toronto in 1897 and 1924, South Africa in 1905 and 1929, Winnipeg in 1909, and Australia in 1914) is not discussed in this chapter. See Dubow (2000) for work on the South African meetings.

References

Collins, H. M., & Evans, R. (2002). The third wave of science studies: Studies of expertise and experience. *Social Studies of Science, 32*, 235–296.

Cooter, R., & Pumfrey, S. (1994). Separate spheres and public places: Reflections on the history of science popularization and science in popular culture. *History of Science, 32*, 237–267.

Dalgleish, W. S. (1894). Geography at the British Association, Oxford, August 1894. *Scottish Geographical Magazine, 10*, 463–473.

Dierig, S., Lachmund, J., & Mendelsohn, A. J. (2003). Introduction: Toward an urban history of science. *Osiris, 18*, 1–19.

Dubow, S. (2000). A commonwealth of science: The British Association in South Africa, 1905 and 1929. In S. Dubow (Ed.), *Science and society in southern Africa* (pp. 66–99). Manchester: Manchester University Press.

Finnegan, D. (2008). The spatial turn: Geographical approaches in the history of science. *Journal of the History of Biology, 41*, 369–388.

Fyfe, A., & Lightman, B. (2007a). Science in the marketplace: An introduction. In A. Fyfe & B. Lightman (Eds.), *Science in the marketplace: Nineteenth-century sites and experiences* (pp. 1–19). Chicago: University of Chicago Press.

Fyfe, A., & Lightman, B. (2007b). *Science in the marketplace: Nineteenth-century sites and experiences*. Chicago: University of Chicago Press.

Gieryn, T. (2006). The city as truth-spot: Laboratories and field-sites in urban studies. *Social Studies of Science, 36*, 5–38.

Gooday, G. (2007). Illuminating the expert-consumer relationship in domestic electricity. In A. Fyfe & B. Lightman (Eds.), *Science in the marketplace: Nineteenth-century sites and experiences* (pp. 231–268). Chicago: University of Chicago Press.

Gregory, J., & Miller, S. (1998). *Science in public: Communication, culture, and credibility*. New York: Plenum Trade.

Higgitt, R., & Withers, C. W. J. (2008). Science and sociability: Women as audience at the British Association for the Advancement of Science, 1831–1901. *Isis, 99*, 1–27.

Keighren, I. M. (2006). Bringing geography to the book: Charting the reception of *Influences of geographic environment*. *Transactions of the Institute of British Geographers, New Series, 31*, 525–540.

Kohler, R. E. (2002). *Landscapes and labscapes: Exploring the lab-field border in biology*. Chicago: University of Chicago Press.

Lightman, B. (2007). *Victorian popularizers of science: Designing nature for new audiences*. Chicago: University of Chicago Press.

Livingstone, D. N. (1995). The spaces of knowledge: Contributions towards a historical geography of science. *Environment and Planning D: Society and Space, 13*, 5–34.

Livingstone, D. N. (2003). *Putting science in its place: Geographies of scientific knowledge*. Chicago: University of Chicago Press.

Livingstone, D. N. (2007). Science, site and speech: Scientific knowledge and the spaces of rhetoric. *History of the Human Sciences, 20*, 71–98.

MacLeod, R. M., & Collins, p. (Eds.) (1981). *The parliament of science: The British Association for the Advancement of Science, 1831–1981*. Northwood: Science Reviews.

Mill, H. R. (1887). Report to council. *Scottish Geographical Magazine, 3*, 521–530.

Miskell, L. (2003). The making of a new 'Welsh metropolis': Science, leisure and industry in early nineteenth-century Swansea. *History, 88*, 32–52.

Morrell, J. B., & Thackray, A. (1981). *Gentlemen of science: Early years of the British Association for the Advancement of Science.* Oxford: Clarendon Press.

Morrell, J. B., & Thackray, A. (Eds.). (1984). *Gentlemen of science: Early correspondence of the British Association for the Advancement of Science.* Oxford: Clarendon Press.

Naylor, S. (2005). Introduction: Historical geographies of science—Places, contexts, cartographies. *British Journal for the History of Science, 38*, 1–12.

Powell, R. C. (2007). Geographies of science: Histories, localities, practices, futures. *Progress in Human Geography, 31*, 309–329.

Rupke, N. A. (1999). A geography of enlightenment: The critical reception of Alexander von Humboldt's Mexico work. In D. N. Livingstone & C. W. J. Withers (Eds.), *Geography and enlightenment* (pp. 319–340). Chicago: University of Chicago Press.

Secord, J. A. (2000). *Victorian sensation: The extraordinary publication, reception, and secret authorship of* Vestiges of the Natural History of Creation. Chicago: University of Chicago Press.

Secord, J. A. (2004). Knowledge in transit. *Isis, 95*, 654–672.

Secord, J. A. (2007). How scientific conversation became shop talk. *Transactions of the Royal Historical Society, Sixth Series, 17*, 129–156.

Shapin, S. (1990). Science and the public. In R. C. Olby, G. N. Cantor, J. R. R. Christie, & M. J. S. Hodge (Eds.), *Companion to the history of modern science* (pp. 990–1007). London: Routledge.

Shapin, S. (1998). Placing the view from nowhere: Historical and sociological problems in the location of science. *Transactions of the Institute of British Geographers, New Series, 23*, 5–12.

Shipton, H. (1892). *August episodes: Studies in sociability and science.* London: A. D. Innes.

Shteir, A. (2007). Sensitive, bashful, and chaste? Articulating the *Mimosa* in science. In A. Fyfe & B. Lightman (Eds.), *Science in the marketplace: Nineteenth-century sites and experiences* (pp. 169–196). Chicago: University of Chicago Press.

Smith, C., & Agar, J. (Eds.). (1998). *Making space for science: Territorial themes in the shaping of knowledge.* Basingstoke: Macmillan.

Sorrenson, R. (1996). The ship as a scientific instrument in the eighteenth century. *Osiris, 11*, 221–236.

Spary, E. (2000). *Utopia's gardens: French natural history from old regime to revolution.* Chicago: University of Chicago Press.

The Meeting (1911). The Meeting of the British Association. *Scottish Geographical Magazine, 27*, 516–531.

Withers, C. W. J. (2001). *Geography, science and national identity: Scotland since 1520.* Cambridge: Cambridge University Press.

Withers, C. W. J. (2002). The geography of scientific knowledge. In N. A. Rupke (Ed.), *Göttingen and the development of the natural sciences* (pp. 9–18). Göttingen: Wallstein.

Withers, C. W. J., Finnegan, D. A., & Higgitt, R. (2006). Geographies other histories? Geography and science in the British association for the advancement of science. *Transactions of the Institute of British Geographers, New Series, 31*, 433–451.

Withers, C. W. J., Higgitt, R., & Finnegan, D. A. (2008). Historical geographies of provincial science: Themes in the setting and reception of the British Association for the Advancement of Science, 1831–c.1939. *British Journal for the History of Science, 41*, 385–415.

Worboys, M. (1981). The British Association and empire: Science and social imperialism, 1880–1940. In R. MacLeod & P. Collins (Eds.), *The parliament of science* (pp. 170–188). Northwood: Science Reviews.

Yearley, S. (2005). *Making sense of science: Understanding the social study of science.* London: Sage.

Testing Times: Experimental Counter-Conduct in Interwar Germany

Alexander Vasudevan

Writing at the end of their now classic study on the experimental landscape of Restoration England, Steven Shapin and Simon Schaffer reflect on what they believe to be "the origins of a relationship between our knowledge and our polity that has, in its fundamentals, lasted for three centuries" (Shapin & Shaffer, 1985, p. 342). It is, they insist, "far from original to notice an intimate and an important relationship between the form of life of experimental natural science and the political forms of liberal and pluralistic societies" (p. 342). But as they go on to suggest, "we are no longer so sure that the traditional characterization of how science proceeds adequately describes its reality, just as we have come increasingly to doubt whether liberal rhetoric corresponds to the real nature of the society in which we now live" (p. 343). Although written over 20 years ago, their remarks still seem on the surface to be tellingly apposite. I have to confess, however, that I have always found their argument to be puzzling and historically questionable. It is, on the one hand, far too easy to see in our current "Age of Emergency" a recasting of the settlement between scientific knowledge production and social ordering. There is, after all, a large body of work that has begun to examine the unseemly affinities between experimental matters of fact, biopolitical modes of rationality, and contemporary forms of neoliberal governance (Cooper, 2008; Ong & Collier, 2004; Rajan, 2006; Rose, 2006). On the other hand, although there is much to recommend this view, my own aim in this chapter is to reflect on the *historicity* of our own experimental turn. I wish to focus attention on the "historical background noise" to a rather different, albeit complementary, understanding of the alignment of experimental labor and political calculation. Specifically, I concentrate on in this case the new experimental arrangements mobilized by the psychiatric community in Germany in the interwar years and the various contemporaneous attempts to counter the credibility of the psychiatric sciences.

A. Vasudevan (✉)
School of Geography, University of Nottingham, Nottingham NG7 2RD, UK
e-mail: alexander.vasudevan@nottingham.ac.uk

P. Meusburger et al. (eds.), *Geographies of Science*, Knowledge and Space 3,
DOI 10.1007/978-90-481-8611-2_11, © Springer Science+Business Media B.V. 2010

Such a view builds on an important and increasingly accepted theme among historians of twentieth-century Germany, namely, that the history of modern Germany is above all the history of a "particular national variant of biopolitics" (Dickenson, 2004, p. 3), understood here as the accreted set of discourses, practices, and institutions focused on the *experimental* care and improvement of individual bodies and the collective body of national populations (see also Crew, 1998; Eghigian, 2000; Kaufmann, 1999). The literature on the emerging master narrative of biopolitical modernity is indeed immense, and it is not my intention to revisit the evidential particulars of such debates (Engstrom, 2004; Lerner, 2003; Roelcke, 2001). Suffice it to say for our purposes that the psychiatric sciences at issue here had themselves emerged from the historically distinctive set of terms by which development of the German *Sozialstaat* (welfare state) during the long nineteenth century was characterized and in which responsibility for interpreting and regulating "the social" was increasingly handed over to the psychiatric community.

In an earlier study of psychotechnical testing in Berlin (Vasudevan, 2006), I attempted to retrace a whole series of experimental embodiments scientifically fashioned and tested in the psychiatric laboratories of Weimar Germany, practices that were primarily tasked with combating the impact of the modern urban experience. Building on recent approaches to the problem of scientific experimentation (Collins, 1992; Galison, 1987, 1997; Gooding, Pinch, & Schaffer, 1989; Schaffer, 1997; Shapin & Schaffer, 1985; Sibum, 1995), I was able to show to what extent the uncertainties of the Weimar metropolitan experience—down to the level of posture, habit, and gesture—were increasingly subject to the exigencies of scientific manipulation and experimental scrutiny. As a result, I was ultimately able to make the case that the rise of psychological expertise during the period of classical German modernity (1890–1930) played "a key role in providing the vocabulary, the information, and the regulatory techniques for the government of individuals" (Rose, 1999, p. 103).

My intention, in this chapter is to open dimensions of past "experimental" practice that I was unable to consider fully in that earlier study. Scientific experimentation, as Galison (1997) once suggested, is

> not captured by procedure alone, even the expanded sense of procedure that includes the protocol-escaping *Fingerspitzengefühl* or tacit knowledge that has so captivated commentators from Polanyi to more recent practitioners of social studies of knowledge. Beyond bench skills, experimentation draws on and alters broader cultural values [and vice versa]. (p. 62)

With this in mind, I wish to shift attention away from a discussion of controlled laboratory arrangements and embodied scientific protocols to an alternative set of experimental systems (see Rheinberger, 1997) that seized on and countered the credibility of psychiatric expertise. To do so, I wish to focus on a series of modernist experiments in interwar Germany that were not only influenced by the experimental life sciences, they actively reconfigured psychiatric science—its epistemic assumptions, its practical arrangements—as a series of critical aesthetic interventions that were themselves tasked with performing scientific experiments with their own alternative regimes of truth. I am drawing implicitly, in this regard, on the work of Michel Foucault, who in the recently published Collège de France lecture course on "La Naissance de la Biopolitique," introduced the concept of *contre-conduit* (counter-conduct) in order to account for an array of tactics whose objective, as he understood it, was to "refuse and contest the governmental calculations of the state in its various manifestations" (Foucault, 2004, p. 363).[1]

My methodological ambition is therefore to set modernist and psychiatric experimentation to resonate with one another, not in order to reaffirm the biopolitical arrangements of the Weimar metropolis but rather to expand and recast the understanding of oppositional experimental communities. Indeed, "activism from below" (Goldberg, 2002, p. 31) guaranteed that the everyday life of modern German psychiatry was not able to provide a simple conduit for therapeutic initiatives and did not make patients the passive objects of superimposed rationalities (Eghigian, 2000). Dissension had numerous points of application, and from a German perspective it undeniably spoke to a variegated microgeography ranging from nineteenth-century opposition to the professionalization of the psychological sciences to the litigious everyday environment of the Wilhelmine and Weimar *Sozialstaat* (see Crew, 1998; Dickenson, 1996; Engstrom, 2004; Goldberg, 1999, 2003; Hong, 1998). As important as these examples are, my own aim is to sketch something of the epistemological field (or experimental counter-space) that was shared between new psychiatric theories and modernist art in interwar Germany, particularly Berlin. A major desideratum of this study is to therefore move toward a more flexible "analysis of the *actual dynamics of cultural interaction* between aesthetic and psychological modernism" (Micale, 2004, p. 17; see also Gilman, 1993; Gordon, 2001; Ryan, 1991; Sass, 1992).

In this chapter, I focus on two interlocking case studies which, taken together, retrace the multiple traffickings between psychiatric experimentation and modernist art. In the first study, I revisit the traumatic reenactments of Berlin Dada in the broader context of industrialized war, rationalized work, and metropolitan life that characterized the Weimar era. As I seek to show, Dada montage pantomimed psychic conflicts and traumas not only to critically reflect on their purported psychiatric effects but to turn these effects back on the very social order that created them (Foster, 2004, p. 156; Sadoff, 1998, p. 7). The second study explores the psychotechnical techniques that were so crucial to the assumptions and operations of Brechtian epic theater as it developed during the late 1920s and early 1930s. The point is not only to query the place of experimentation (or the space of knowledge formation) but also to reconsider *what* may properly count as "experimental" knowledge in the first place. Whereas "spatialized approaches" to the "making and maintaining" of scientific knowledge have undoubtedly magnified our understanding of scientific practice, the cultural dimensions of experimentation have garnered less attention within geography (Smith & Agar, 1998, p. 3; see Matless & Cameron, 2006). The two case studies charted in this chapter seek to redress this omission by retracing the different ways in which notions of "experiment" became an organizing grammar for the Weimar avant-garde. Whether it be performance art or political theater, each offers, as I propose to show, a critical perspective on the extension of the experimental into nonscientific zones. Indeed, it is only by moving beyond the confines of the laboratory that the task of mapping the cultural geography of Weimar science can really begin (see Kohler, 2002).

Traumatic Territories

The origins of this study can be traced back to a Dada soirée that took place in Berlin at the Meistersaal on Köthenerstrasse 38 in May 1919 (see Bergius, 2000). The soirée was accompanied by the following newspaper announcement in the *Neue Berliner 12 Uhr Zeitung*: "The Dadaists differ from many who have [taken the stage] since August 1914 primarily in that they only simulate insanity" (May 20, 1919). Whether this statement is directly attributable to one of the Berlin Dadaists remains unclear. What is certain, however, given the announcement's allusion to August 1914, is that the "simulation" with which the Dadaists were primarily preoccupied was that of war neurosis, in other words, the condition more commonly referred to as "shell shock." Cases of shell shock reached epidemic proportions during World War I as soldiers, forced to confront the experience of mass industrialized warfare, returned from the front in their thousands, suffering from sleeplessness, uncontrollable shaking, and psychogenic disorders of speech, sight, hearing, and gait (see Lerner, 1996, 2000, 2001). It was these very cases of war neurosis that were subsequently to furnish Berlin Dada with an experimentally sanctioned repertoire of physical and psychic symptoms, though they also heralded a literal displacement of trauma and its treatment from the trenches and clinics to the sites and venues of postwar metropolitan culture.[2] It was precisely this restoration of traumatic behavior that was seized on by the Berlin Dadaists as they fashioned their own "laboratory" out of the city's cabarets, galleries, and revues.

As Grosz (1955/1986) would later write, "I could write page after page about [World War I], but everything I could say about it is already in my drawings" (p. 102). Grosz had himself enlisted for combat in November 1914, though rather than being sent to the front, he was soon committed to a *Lazarett*, or military hospital, and in May 1915 was declared unfit for service. Grosz was called up again in early 1917 only to be rehospitalized. He spent 4 months in two separate mental clinics, both in the vicinity of Berlin. Like Grosz, John Heartfield also spent time in a mental hospital in 1915 after a "simulated" breakdown before he was dispatched to the front. When he was finally discharged, he took up a temporary position as an "unreliable" postman in the Grunewald district in Berlin (Doherty, 1997, pp. 93, 102; Herzfelde, 1962). Unlike Grosz and Heartfield, Richard Huelsenbeck spent much of the war in Zurich where he came into contact with the coterie of artists who would form the International Dada movement. Himself a founding member of Berlin Dada, Huelsenbeck went on to become a neuropsychiatrist in Berlin in the 1920s, interning with Karl Bonhoeffer at the Charité hospital. He would later practice Jungian psychoanalysis in the United States under the name Charles R. Hulbeck.

Of course, any biographical reading of subsequent Dada compositions is purely speculative, though I do want to dwell on Doherty's (1996) assertion that "Berlin Dada engaged very specifically with important aspects of German modernity" (p. xxv; see also Doherty, 1997). For Doherty, montage represented a visual form predominantly conceived to engage with the embodiment of modernity in subjects transformed by industrialized war, rationalized work, and metropolitan life. Such practice was most obviously dramatized in the modest exhibition that was held in 1920 in two ground-floor rooms of a five-story Gründerzeit building at 13 Lützowufer, Berlin. Between June 30 and August 25, the Dr. Otto Burchard Gallery hosted an exhibition dedicated to the First International Dada Fair, a mixed-media show of roughly two hundred objects that included the work of Richard Huelsenbeck, Raoul Hausmann, Hannah Höch, George Grosz, John Heartfield, and Otto Dix. Recent art-historical discussions of the exhibition have increasingly focused on the montage principles to which the exhibition primarily adhered. As Doherty (2003) has argued, Berlin Dada "produced an art of failed revolution" (p. 92) in which the destructive energy as well as the actual collapse of the November Revolution on the streets of Berlin were figured, quite literally, within the formal conventions of Dada montage (see also Doherty, 1997). In using, among other things, photographs, scissors, and glue rather than brushes and paint, the Dadaists were also able to distance themselves from the traditional practices of art-making.

In their self-professed capacity as "monteurs," "industrial machinists," and "technical experts," the Berlin Dadaists, like many of their counterparts in the Weimar avant-garde, elevated the notion of expertise in an effort to attend to the labor of the "artist" in modernity (Schwartz, 2000, p. 405). The Dadaists were admittedly hyperbolic in their "anti-art" diatribes, though these are better read as a critique of the technical procedures of modernism rather than their rejection *tout court*. As Dickerman (2003) has suggested, "Dadaism is about making, about producing art in changed historical circumstances" (p. 8). The accent on "making" is reflected in the materiality of the techniques around which Dada developed as a movement: photomontages, sculptural assemblages, puppets, and typographic experiments. But more importantly, this focus was also a caustic engagement with the materiality of those "changed historical circumstances" to which the Dadaists directed their negative energies. If Dada was, in this way, able to cohere around what Benjamin (1972–1989) once referred to in 1921 as a form of "negative expressionism" (vol. 6, p. 132), the work of Berlin Dada was ultimately preoccupied with the traumatic working through of the unprecedented and senseless slaughter of World War I. War trauma, as I have already intimated, must be counted among the "most basic concerns of Berlin Dada" (Doherty, 1997, p. 88).

As such, the link between trauma and Berlin Dada's new institutional home figures prominently in the movement's founding manifesto, its most important passage highlighting Dada's fixation with shock and dismemberment:

> The highest art will be that which in its conscious content presents the thousandfold
> problems of the day, an art that one can see has let itself be thrown by last week's explosions,
> that is forever gathering up its limbs after yesterday's crash. The best and most extraordi-
> nary artists will be those who every hour snatch the tatters of their bodies out of the frenzied
> cataract of life[.] (Huelsenbeck, 1920/1987, p. 36).

References to "last week's explosions" and "yesterday's crash" make it clear that, for the Dadaists, an experimental encounter with the "shock" of modernity was of a particular time and place, namely, Berlin, 1920. Such references also heralded an unambiguous return to the original scene of war neurosis, which undoubtedly should be seen as a shell-shocked mimesis *in extremis* of those who had quite literally been thrown and traumatized by "explosions." But as a parodic set of strategies, Dadaism was not nihilistic insomuch as it was defensive and immunological. After all, the Dadaist was modeled, as Hugo Ball, one of the movement's founders pointed out, to a much lesser degree on the absolute anarchist than on the "perfect psychologist [who] has the. . . power to shock or soothe with one and the same topic" (Ball [1916] as quoted in Foster, 2003, p. 170).

That the unfolding of trauma ultimately found a high degree of concentration and theatrical expansion with the artistic practices of Berlin Dada squares up, I real-ize, with long-standing approaches to the relationship between mental illness and modernist art (see Felman, 2003; Sass, 1992; Thiher, 1999). However, unlike these earlier studies, this chapter is neither intended to introduce yet another collection of psychoanalytical readings of key cultural texts nor does it offer a series of psy-chobiographical vignettes on the mental lives of Modernist novelists, dramatists, and painters (Micale, 2004, p. 3). In focusing on the "madness" of Dada, one needs to be mindful, if anything, of the fact that the Dadaists were not only reacting to the traumatic strictures of militarized warfare. They were themselves also implicated within a wider set of somatic and kinesthetic vocabularies experimentally devel-oped to sense and adapt to the shock of urban industrial modernity. If it was modern city life that was largely held responsible for the advance of debilitating mental and physical diseases, psychiatric expertise offered, so it seemed, the possibility of mas-tering otherwise diffuse social, political, and cultural disorientations. In the words of Santner (1995), it was as though "scientific or medical knowledge could become the source of a renewed sense of social and cultural location, a sense of certainty as to one's place in a symbolic network" (p. 8). In the context of the Dada Fair, the extent to which the credibility of these clinical assumptions was itself questionable is, as I see it, clearly on display. "We were," as Hausmann (1977) later noted, "in the habit of speaking psychoanalytical gibberish" (p. 6).

However extraordinary and hyperbolic these assertions may seem to be, they are undoubtedly meant to highlight the nature of the "insights" in evidence at the Fair—Dada's "operating theater" *in concreto*. As Bergius (2000) has shown, the exhibition itself was organized to mimetically reproduce the montage logic that inhered in the individual works that cluttered the walls of the gallery. The disorientating surfeit of works on offer was, in this way, made to recall the bombardment of moving images and texts—the posters, electric advertisements, and filmic images—that had already transformed the public spaces of Weimar Berlin while competing for the attention of the city dweller. It is, moreover, here that the inflated relationship between the fragmented montage materializations of Berlin Dada and the bodily materializations of the traumatic psychic shock of war neuroses is most decisively evident (see Doherty, 1997, p. 84). Take, for example, the famous publicity photos from the fair. Hung from the ceiling is Rudolf Schlichter's *Prussian Archangel*, a stuffed officer's uniform fitted with a plastered pig head. On one side, a tailor's dummy has been installed with a light bulb for a head, a metal pole for a leg, and numerous military decorations glued to its torso, undoubtedly a reference to the legion of *Kriegskrüppel* (crippled veterans) and *Prosthetiker* (veterans fitted with prostheses) whose public presence on the streets of Berlin became a cause of growing concern to the state. In fact, just as the Berlin Dadaists were organizing their own exhibition, one of the main clinics for the experimental testing and production of working prostheses, the Oskar-Helene-Heim Klinik in Berlin-Dahlem, was itself receiving visitors to see its display of the latest prosthetic aides. A museum would later open to the public in 1921, with the stated aim of "representing and ordering [the] most unaesthetic material in the world such that the impression it makes is at least not too unaesthetic" (Biesalski, n.d.).

What I have been building up to proposing, then, is that the Dadaists themselves offered perhaps one of the best characterizations of the traumatizing effects of a historically specific mode of experimental knowledge—the psychiatric regulation of war neuroses. This is a provocative claim though the burden of evidence suggests, in my view, a necessary point of mediation between Dadaist aesthetics and psychological experimentation. Consider *Die Gesundbeter: The Faith-Healers or Fit for Active Service*, an early lithograph by Grosz in which the clinical refashioning of the damaged and traumatized worker–soldier is pushed to the point of parodic excess. Republished in the 1920 series, *Gott mit Uns*, which was itself included in the Fair, the image shows an army doctor examining the corpse of a soldier with an ear trumpet. He pronounces the corpse *kriegsverwendungsfähig* (KV), literally meaning "capable of war use," in other words, fit for active service. The French subtitle to the lithograph, *Le triomphe des sciences exactes* (The triumph of scientific precision), only reinforces Grosz's critique.

The degree to which Berlin Dada offered a riposte to the military-industrial experience is further underlined by the publicity photos of the exhibition. In the main photo of the exhibition, the two largest paintings of the show are clearly visible. On the left-hand side is Otto Dix's now lost *War Cripples* (45% *Employable*), with its procession of maimed veterans, one shaking wildly, clearly a victim of traumatic neurosis. In the background is Grosz's *Germany, A Winter's Tale*, its use of counterposed lines and violently arranged curves and angles evoking the turbulent events taking place on the streets of Berlin. A further publicity photo shifted attention to a series of large-scale posters, verbal experiments, and sound poetry in striking boldface typography, perhaps highlighting Benjamin's (1979) assertion that the "typographical experiments undertaken by the Dadaists stemmed. . . not from constructive principles but from the precise nervous energy of these literati" (p. 61).

In the end, reviews of the Dada Fair were unfailing in their criticisms of the exhibition's methods of "reflected negation" (Sloterdijk, 1987, p. 397). Even Kurt

Tucholsky, one of the exhibition's more sympathetic and well-meaning reviewers, was quick to identify Dada's negative *modus operandi*: "I know exactly what these people want: The world is motley, senseless, pretentious, and intellectually inflated. They want to despise, show up, deny, destroy that" (Tucholsky, 1920). Another review published in the *Tägliche Rundschau* in the summer of 1920 by Werner Leibbrand, a doctor at Berlin's famous Charité hospital, went so far as to identify several formal qualities shared by Berlin Dada and art made by his patients suffering from mental illness:

> A red thread runs through all of them. . . In the mental ward of the Charité, we have a little exhibition of such works. And now we have [found] in the Dadaist artworks a borderline case of a symptom complex somewhere between an organic mental illness and a more generalized psychopathology, and so we psychiatrists are grateful to the Dadaists for their exhibition. Whether one does right by the people by letting them see it, is questionable: the collection belongs in the Charité. (Leibbrand, as quoted in Doherty, 1997, p. 91).

In one sense, Dr. Leibbrand was only revisiting the widespread clinical revaluation of the art of the mentally ill in the early twentieth century, an assessment to which the work of the German psychiatrist and art historian Hans Prinzhorn (1922) was of particular epistemic significance, especially given his reception by various modernists (Foster, 2004, pp. 194–195). The doctor's "diagnosis" also recalled a groundswell of jeremiads among German mental health professionals in the revolutionary aftermath of World War I (Micale, 2004, p. 3). Whereas some doctors identified the onset of the revolution with a mood "conducive to the lifting and waning of hysterical symptoms in soldiers," the majority explicitly blamed Germany's military defeat and the toppling of the Wilhelmine order on the exhaustion of the nation's nerves (Singer, 1919, p. 331). Other physicians argued that the Revolution provided incontrovertible scientific evidence of the psychogenic nature of political activism, not to mention Dada's singular capacity for political agitation. In response to revolutionary activities, several prominent doctors, including Eugen Kahn and Kurt Hildebrandt even published case histories of revolutionary actors, material replete with clinical typologies of different psychopathological "classes" (see Kahn, 1919, for example). The Hamburg psychiatrist Professor Wilhelm Weygandt promoted his growing collection of the art of the mentally ill for use as defamatory evidence against the modernist aesthetics of the postwar avant-garde—the futurists, Dadaists, and members of the Bauhaus—all of whose work was deemed irredeemably degenerate, demented, and schizophrenic.

Such perorations notwithstanding, I think it is clear that, in experimenting with the disjunctive form and aggressive content of montage, the Berlin Dadaists were able to "reanimate" and "embody" the alienating experiences of modern metropolitan life and war trauma. Not only did the production of Dada montage lend itself to the articulation of traumatic shock and give concrete form to a dismembered, emasculated male body, but its reception by a beholder issued its own traumatic shock. "The point," writes Dickerman (2003), "[was] not so much virtuoso object production, although this [was] at times a consequence, but intervention and activation of the terrain of modern culture itself" (pp. 9–10).

Gestural Polities

The scandalous montage practices used by the Berlin Dadaists to demystify the circumstances of traumatic bodily shock were a source of widespread interest to members of the Weimar avant-garde (see Lavin, 1993; Leslie, 2002; Möbius, 2000). Benjamin, for one, likened the perceptual modality initiated by Berlin Dada to the production and reception of film. "Dadaism," he wrote, "attempted to produce with the means of painting (or literature) the effects which the public today seeks in film" (Benjamin, 1996–2003, vol. 3, p. 118). And yet, although Benjamin may have seen film, with its montage techniques, close-ups, and variable framings, as the medium most adequate to the shock sensations of modern metropolitan life, his engagement with Brechtian dramaturgy arguably came to occupy an equally prominent place within the development of his aesthetic theory. As in the case of Berlin Dada, Benjamin singled out montage as one of the principle techniques underwriting the practice of Brecht's epic theater.

As important as the affinities between Benjamin and Brecht undoubtedly are, I wish to push the argument in another direction (see Doherty, 2000; Giles, 1998; Hansen, 1999, 2004). Specifically, I concentrate on the techniques of experimental psychiatry that both Benjamin and Brecht understood to be so crucial to the assumptions and operations of epic theater as it evolved during the late 1920s and early 1930s. Just as the "crudities" and "excesses" of Dada found a supporting "evidential context" (Benjamin, 1996–2003, vol. 3, p. 118; Schaffer, 1992, p. 330) amidst the strictures of the psychiatric sciences, I wish to demonstrate to what extent the work of Brecht was itself indebted to the experimental arrangements of German industrial psychology. But unlike Dada's multifarious attempts to contest the experimental dissimulation of war trauma that centered on the excess of gesture and representation that was its defining symptom, Bertolt Brecht's contemporary experiments in dramaturgy were characterized by a rather different "economy of gesture," namely, scientific management's minimum gesture and psychotechnical testing's controlled gesture (McCarren, 2003, p. 6).The story here is not of extreme physical shock effects but indeed the reverse, as "gestural knowledge" was placed in the service of psychiatric scrutiny and experimentation (Sibum, 1995, p. 76). Of course, the experimental testing methods of occupational counseling (*Psychotechnik*) were not limited to the development of "measures for economizing energy and improving [workplace] efficiency" (Killen, 2003, p. 215). Although such gestural techniques became central to the aesthetics of a range of avant-garde modernisms (see McCarren, 1998, 2003), my own aim is to concentrate on the repertoire of practices that became crucial to the theory and practice of Brecht's epic theater.

In this context, Brecht's 1926 comedy *Mann ist Mann* offers a particularly compelling case study of the historically significant connections between gestural knowledge in experimental psychiatry and epic theater. The play recounts the "reverse engineering" of Galy Gay, a benign Irish packer, into a "ruthless British soldier in a colonial machine-gun regiment" (Doherty, 2000, p. 450). The very techniques that transformed Gay's *physis* over the course of the play were, I contend, tantamount to the practice of Brecht's epic theater, ranging from his writings collectively known as *Versuche* (experiments) to his subsequent theatrical oeuvre, including important works such as the *Badener Lehrstücke*, *Aufstieg und Fall der Stadt Mahagonny*, and *Die Dreigroschenoper*.

The development of *Mann ist Mann* can be traced back to sketches for a "Galgei project" penned while Brecht was serving as a military medical orderly in a mental hospital (*Nervenklinik*) near Augsburg in 1918. Brecht returned to the project in the spring and summer of 1920 and in the mid-1920s when he had finally settled in Berlin. Early notes for the play indicated that Galgei was originally set in a civilian Augsburg. It was only after Brecht's move to Berlin that he reworked the original draft manuscript with the help of Elizabeth Hauptmann and changed the setting to colonial India. The main character, Galgei, a Bavarian carpenter, became Galy Gay, and the play became *Mann ist Mann*, ostensibly Brecht's canonical experiment within the conventions of epic theater. The play premiered simultaneously in Darmstadt and Düsseldorf on September 25, 1926. Brecht would later revise the play for an unsuccessful run of performances that he himself directed at Berlin's Staatliches Schauspielhaus in February 1931 (Doherty, 2000, pp. 450–451; see Brecht, 1996, vol. 2, pp. 406–423).

For Brecht, the staging of a new "epic" style promised an alternative to the tenets of Aristotelian drama with their inevitable plotting of "the hero into situations where he [revealed] his innermost being" (Brooker, 1994, p. 187). Epic theater was dedicated to the foreclosure of spectatorial empathy and, with it, an audience's identification with a play's principal protagonist, not to mention their surrender to the "suspense and consolations of the well-made play" (p. 188). Brecht associated these conventions with the categories of "mimesis" or "catharsis" and the anaesthetizing of particular emotions (p. 188). Epic theater, as explained by Brecht (1964/2000),

> does not make the hero the victim of an inevitable fate, nor does it wish to make the spectator the victim, so to speak, of a hypnotic experience in the theatre. In fact, it has as a purpose the 'teaching' of the spectator a certain quite practical attitude... while he [*sic*] is in the theatre. (p. 78).

If the epic stage was, in other words, to be suitably instructive and *useful*, it had to effectively transform its audience from "a mass of hypnotized test subjects" into "a theater full of experts" (Brecht, 1996, vol. 24, p. 59).

This task was accomplished through the conspicuous use of stage machinery, films, placards, emblematic props, and music set in a deliberately montaged sequence. Only in this way could the new narrative content signaled by the term *epic* be expected to instruct in a "dialectical, non-illusionist, and non-linear manner, declaring its own artifice as it hoped also to reveal the workings of ideology" (Brooker, 1994, p. 191). As Brecht (1996, vol. 23) noted, "the exposition of the story and its communication by suitable means of alienation [*Verfremdung*] constitute the main business of the theater" (p. 81). Such alienations, he added, are "designed to free socially conditioned phenomena from that stamp of familiarity that protects them against our grasp today" (p. 81). Achieving these practical effects and forms of *Verfremdung* depended, according to Brecht, on "treating society as though all of its actions were performed as experiments" (p. 85). "The new school of drama must incorporate the 'experiment' into its form," he proclaimed (vol. 24, p. 67), a point echoed by Adorno (2001), who remarked that Brechtian dramaturgy was in fact governed "by a kind of time-space, an experimental time which more closely resembles that of the repeatable 'laboratory experiment' than it does the time of history" (p. 75). Benjamin (1977) similarly highlighted Brecht's preoccupation with rehearsing, in his words, "elements of reality as though it were setting up an experiment" (p. 4).

Of course, the accent on the machinery of "experimental philosophy" with its own technologies of accreditation was not arbitrary. Nor did it presuppose a simple circuit of influence that ran in a straight line from the purportedly "credible" realm of science to certain theatrical rituals. By the late 1920s, Brecht had become interested in the findings of behavioral psychology and the development of logical empiricism, which led him to regularly attend lectures in 1932–1933 at the Berlin branch of the Gesellschaft für empirische Philosophie (Society for Empirical Philosophy). Brecht was a close reader of *Erkenntnis*, the flagship journal of the logical empiricists, and his own annotated copies offer an important point of purchase on his idiosyncratic reading of logical empiricist thinking and its relationship to the nascent field of behavioral psychology (see Danneberg & Müller, 1990; Giles, 1998; Knopf, 1974). Though the full implications of Brecht's interest in logical empiricism and behaviorism are beyond the compass of this chapter, an interest in plotting the "human psyche's outward effects" (Brecht, 1964/2000, p. 67) was certainly crucial to his mode of theatrical experimentation. As he eventually suggested, "psychology is an important field for the dramatist" (Brecht, 1996, vol. 22, pt. 1, p. 113), and in *Mann ist Mann* he explores this conviction by adapting the methods of psychotechnics and Taylorist industrial management, which had become a veritable craze within the metropolitan centers of Weimar Germany (see Meskill, 2004; Nolan, 1994; Rabinbach, 1990).

Having emerged out of the morass of wartime psychiatry, these techniques became the tools for monitoring within a controlled laboratory milieu the injurious everyday effects of the modern metropolis. Used originally to prepare returning veterans for various occupations, including streetcar conductors, typesetters, and plant engineers, such experimental methods were widely used in the training of the "salaried masses" (Kracauer, 1930/1998) that formed the core of Weimar Berlin's anomic, bureaucratized society. In fact, by the early 1920s, no fewer than 170 psychotechnical testing stations had sprung up in Germany, and more than 60 firms had established their own stations offering what Adorno (1951/1974) later described as "the consummation of the division of labour within the individual" (p. 231). In practical terms, testing divided each minute component of the work process into individual mental and sensorimotor skills in order to facilitate the calibration of physical and political economy. After all, as Walter Benjamin opined in a 1930 radio broadcast, "in what realm of life are habits more easily formed, where are they more vigorous, where do they more fully encompass entire groups of people, than at work?" For Benjamin, it made perfect sense that psychotechnical experiments should "test the posture of specific occupations entirely apart from the content of the work itself, focusing instead on gesture, aptitude, and capability" (Benjamin, 1972–1989, vol. 2, pt. 2, pp. 670–671).

That *Mann ist Mann* was ultimately dependent upon the same understanding of occupations, habits, postures, and gestures that characterized the psychotechnical testing enterprise is revealed through the crude and violent reverse-engineering of Galy Gay. He is forced, like the subject of the psychotechnical test, to reenact, through a scripted series of experiments and tests, the conduct and gestures of the vocation he is forced to take up (Doherty, 2000, pp. 449–450). "Bit by bit," writes Benjamin (1996–2003) of Gay, "he assumes possessions, thoughts, attitudes, and habits of the kind needed by a soldier in war; he is completely reassembled" (vol. 2, p. 369). Indeed, the extent to which Brecht envisioned the technician or technical worker as a counterpart to Galy Gay is made strikingly manifest in a 1929 notebook entry that Brecht made while undertaking revisions to the play: "*Mann=Mann*/counterpart: the technician/... for the worker is no prince. he [*sic*] comes into being not by birth, but insofar as he is violently remade. therefore [*sic*] all human beings can be turned into workers" (quoted in Doherty, 2000, p. 451).

And yet, as much as Gay was forced to felicitously reproduce the gestic aptitudes of one Jeraiah Jip, a machine-gunner who had gone missing from his regiment, his behavior was also intended to provide an "occasion for exemplary gestic acting" (Doherty, 2000, p. 465). "Observation," Brecht writes in the *Kleines Organon für das Theater*, "is a major part of acting. The actor observes his [*sic*] fellowmen with all his nerves and muscles in an act of imitation. ... To achieve a character rather than a caricature, the actor looks at people as though they were playing him their actions, in other words as though they were advising him to give their actions careful consideration" (Brecht, 1996, vol. 23, p. 86). For Brecht, the technique through which Galy Gay was able to assume the *Haltung* (disposition) of the soldier by mimicking its gestures was really the technique of epic acting itself. Only in this way, so Brecht believed, could the manner of Gay's transformation from "firm characteristics to push-button behaviour patterns" (Adorno, 1951/1974, p. 231) become a device capable of transforming the theater spectator from a subject of psychotechnical testing into a critical participant capable of adjudicating techniques that had earlier seemed embodied, tacit, and irremediably local. "The audience," in Brecht's view, "has got to be a good enough psychologist to make its own sense of the material I put before it" (Brecht, 1964/2000, p. 14).

That these performances experimented with an alternative version of *Psychotechnik* was again a point not lost on Benjamin. He himself noted in 1930 the extent to which such work had already gained "insight into the great system of tests" and "the colossal laboratory of a new science that has quickly established itself in Germany: the science of work" (Benjamin, 1972–1989, vol. 2, pt. 2, pp. 667–668; see Schräge, 2001). Benjamin, one of Brecht's more astute interlocutors, imagined a suitably enlightened audience that had insight into vocational aptitude and performance tests not only through mass-media attention to those technologies but also through direct experience in the laboratory and in a city where one's ability to withstand and accommodate the shocks of metropolitan modernity had increasingly become the ultimate test. Benjamin's (1996–2003) gloss on *Psychotechnik* was thus directed at reversing their original intentions and developing a rather different "form of expertise" through complementary experiments in radio, film, and Brechtian epic theater, experiments that, in Benjamin's words, would be capable of producing "the revolutionary in a test tube" (vol. 2, p. 369).

Despite these political entailments, Brecht's 1931 production of *Mann ist Mann* was hardly popular. It drew sharp criticism from audiences all too intimately ensconced within the experimental cultures of occupational testing. The play, and in particular the epic acting of Peter Lorre as Galy Gay, were greeted with angry shouts and derisive jeers rather than thoughtful critical reflection. Several reviewers even saw Lorre's deliberately staccato rendition of Gay's transformation as neurasthenically susceptible, like Gay's antecedents in the Dada movement, to the effects of shock and terror (see Bab, 1931). Of course, while epic theater lacked credibility and warrant as an experimental "method," Benjamin could still turn to the production and reception of film as an *experimental* counterpart to psychotechnical testing. As he concluded in the first version of the now canonical "Kunstwerk" essay, "the great majority of city dwellers in offices and factories are in the course of the workday expropriated of their humanity" (Benjamin, 1996–2003, vol. 3, p. 111). In the evening, "these same masses fill the cinemas, to witness the film actor taking revenge on their behalf not only by asserting his [*sic*] humanity (or what appears to them as such) against the apparatus, but by placing that apparatus in the service of his triumph" (p. 111).

Concluding Comments

Experimental psychiatry, whether the regulation of war trauma or the management of workplace efficiency, furnished Berlin Dada and Brechtian epic theater with a new repertoire of performance styles and representational techniques. Such scientific experiments also promoted a set of highly serviceable themes seized on by various members of the Weimar avant-garde. And yet, the experimental program promoted by both Berlin Dada and Brecht's epic theater was not of a piece with the evidential mechanisms by which the German psychiatric community attempted to establish a reliable and manageable model for adjudicating the mental well-being of society. Rather than consolidate what counted as proper psychiatric knowledge, their aim, if anything, was to establish an alternative set of experimental matters of fact to those brokered by professional psychiatry (see Shapin & Schaffer, 1985, p. 22). Although aesthetic experimentation of the kind present in Dada and epic theater undoubtedly raised questions about scientific claims to the mastery of modernity's traumas, their own assertions were equally fragile. Experimentation of the kind I have described in the preceding pages was, however, far less successful in making the transit to the wider circuits of public culture. In the end, both Dada and epic theater fashioned "scenes of inquiry" (Jardine, 2000) that were themselves remarkably limited, be they the epic stage or the gallery space.

I am not the first to note the role of the German psychiatric community in producing "regimes of truth" whose claims to validity were imposed by the demands of modernity. Shapin and Schaffer (1985), for example, have already examined ways in which the epistemic authority of Weimar psychiatry was secured through a multitude of technologies that guaranteed that certain solutions to the problem of psychiatric knowledge were, in turn, solutions to the problem of social order (p. 332). In this chapter, however, I have attempted to move beyond the purview of those investigations in order to draw attention to an alternative experimental program that actively contested the theories and practices of mainstream German psychiatry. The combined cultural materials recounted in these pages are thus intended not only as contributions to a historical geography of experimentation but also as a political anatomy and an addition to what Ronell (2003a) has described as an "age of experimentation" (p. 573). In her eyes, one of the defining predicaments of modernity is indeed the "adherence to the imperatives of testing" (p. 563). "Today's world," she writes, "is ruled conceptually by the primacy of testing—nuclear, drug, HIV, admissions, employment, pregnancy, and DNA tests, the SAT, GRE, and MSAT" (p. 563; see also Ronell, 2003b, 2005). She is, of course, painting in broad brushstrokes, and I do take a certain distance from her universalizing form of theorizing about the "modern experimental turn" (2003a, p. 563). Her interest is resolutely philosophical in nature and involves, in particular, "our relation to explanatory and descriptive language, truth, conclusiveness, result, probability, process, and identity" (p. 563). By contrast, I have sought to remain alert to the ways in which the scene of experimentation was itself a contingent historicogeographical matter marked by the intersection of the aesthetic and the psychological, the political and the philosophical. At the same time, I must underscore that this study is not a blithe embellishment of the existing historiography on experimental practice. On the contrary, the point is to show *how* more expansive epistemic cultures and politicointellectual projects promised by the modern experimental turn have already taken on a host of historical forms and achieved varying degrees of local success. To do so is also to heed Ronell's warning (2003a) that

> the experimental turn as we now know it from a history of flukes, successes, and near misses, in its genesis and orientation, travels way beyond good and evil. Its undocumented travel plan—there are so many secret destinations of which we remain ignorant—is perhaps why experimentation is a locus of tremendous ethical anxiety. (p. 563)

Although the two episodes described in this chapter are not intended to offer the single definitive course through Ronell's experimental "geography," they nevertheless strike a path through what remains an uncertain and changing landscape.

Acknowledgment I am grateful to Pion Limited, London, for permission to publish an abridged version of an essay that first appeared as "Experimental Embodiments, Symptomatic Acts: Theatres of Scientific Protest in Interwar Germany," *Environment and Planning A, 39,* (2007), 1812–1837.

Notes

1. Unless otherwise noted, all English translations in this chapter are my own.
2. One of the writers of *The Cabinet of Dr. Caligari* was himself treated for war neurosis (Lerner, 2003, p. 219).

References

Adorno, T. (1974). *Minima Moralia* (E. Jephcott, Trans.). London: Verso (Original work published 1951).

Adorno, T. (2001). The schema of mass culture (N. Walker, Trans.). In J. M. Bernstein (Ed.), *The culture industry: Selected essays on mass culture* (pp. 61–97). London: Routledge.

Bab, J. (1931). Nachtrag zum Falle Brecht [Addendum to the Brecht case]. *Die Hilfe*, *37*, 187–189.

Benjamin, W. (1972–1989). *Gesammelte Schriften* [Collected writings] 7 Vols, R. Tiedemann & H. Schweppenhäuser (Eds.). Frankfurt am Main: Suhrkamp.

Benjamin, W. (Ed.). (1977). What is epic theatre? In A. Bostock (Trans.), *Understanding Brecht* (pp. 1–13). London: Verso.

Benjamin, W. (1979). One-way street. In E. Jephcott & K. Shorter (Trans.), *One-way street and other writings* (pp. 45–104). London: NLB.

Benjamin, W. (1996–2003). *Selected Writings* (M. Bullock & M. Jennings, Eds., R. Livingstone, G. Smith, & E. Jephcott, Trans., 4 Vols). Cambridge, MA: The Belknap Press.

Bergius, H. (2000). *Montage und Metamechanik: Dada Berlin, Artistik von Polaritäten* [Montage and metamechanics: Dada Berlin, artistry of polarities]. Berlin: Mann.

Brecht, B. (1996). *Große kommentierte Berliner und Frankfurter Ausgabe* [Great annotated Berlin and Frankfurt edition], 30 vols. Frankfurt am Main: Suhrkamp.

Brecht, B. (2000). *Brecht on theatre: The development of an aesthetic* (J. Willett, Ed. & Trans.). New York: Hill and Wang (Original translation published 1964).

Brooker, P. (1994). Key words in Brecht's theory and practice of theatre. In P. Thomson & G. Sacks (Eds.), *The Cambridge companion to Brecht* (pp. 185–200). Cambridge, England: Cambridge University Press.

Collins, H. M. (1992). *Changing order: Replication and induction in scientific practice* (2nd ed.). Chicago: University of Chicago Press.

Cooper, M. (2008). *Life as surplus: Biotechnology and capitalism in the neoliberal era*. Seattle: University of Washington Press.

Crew, D. (1998). *Germans on welfare: From Weimar to Hitler*. Oxford: Oxford University Press.

Danneberg, L., & Müller, H.-H. (1990). Brecht and logical positivism. *The Brecht Yearbook*, *15*, 151–163.

Dickenson, E. R. (1996). *The politics of German child welfare from the Empire to the Federal Republic*. Cambridge, MA: Harvard University Press.

Dickenson, E. R. (2004). Biopolitics, fascism, democracy: Some reflections on our discourse about "modernity." *Central European History*, *37*, 1–48.

Dickerman, L. (2003). Dada gambits. *October*, *105*, 4–12.

Doherty, B. (1996). *Berlin Dada: Montage and the embodiment of modernity, 1916–1920*. Unpublished doctoral dissertation, Department of Art History, University of California, Berkeley.

Doherty, B. (1997). The trauma of Dada montage. *Critical Inquiry*, *24*, 81–132.

Doherty, B. (2000). Test and *Gestus* in Brecht and Benjamin. *Modern Language Notes*, *115*, 442–481.

Doherty, B. (2003). The work of art and the problem of politics in Berlin Dada. *October*, *105*, 73–92.

Eghigian, G. (2000). *Making security social: Disability, insurance, and the birth of the social entitlement state in Germany*. Ann Arbor: The University of Michigan Press.

Engstrom, E. (2004). *Clinical psychiatry in imperial Germany: A history of psychiatric practice*. Ithaca, NY: Cornell University Press.

Felman, S. (2003). *Writing and madness* (M. N. Evans, Trans.). Stanford, CA: Stanford University Press.

Foster, H. (2003). Dada mime. *October, 105*, 166–175.

Foster, H. (2004). *Prosthetic gods*. Cambridge, MA: The MIT Press.

Foucault, M. (2004). *La Naissance de la Biopolitique* [The birth of biopolitics]. Paris: Seuil.

Galison, P. (1987). *How experiments end*. Chicago: University of Chicago Press.

Galison, P. (1997). *Image and logic: A material culture of microphysics*. Chicago: University of Chicago Press.

Giles, S. (1998). *Bertolt Brecht and critical theory*. London: Peter Lang.

Gilman, S. (1993). *Freud, race, and gender*. Princeton, NJ: Princeton University Press.

Goldberg, A. (1999). *Sex, religion, and the making of modern madness: The Eberbach asylum and German society, 1815–1849*. Oxford, England: Oxford University Press.

Goldberg, A. (2002). The Mellage trial and the politics of insane asylums in Wilhelmine Germany. *The Journal of Modern History, 74*, 1–32.

Goldberg, A. (2003). A re-invented public: "Lunatics' rights" and bourgeois populism in the Kaiserreich. *German History, 21*, 159–182.

Gooding, D., Pinch, T., & Schaffer, S. (Eds.). (1989). *The uses of experiment*. Cambridge, England: Cambridge University Press.

Gordon, R. B. (2001). *Why the French love Jerry Lewis: From cabaret to early cinema*. Stanford, CA: Stanford University Press.

Grosz, G. (1986). *Ein kleines Ja und ein großes Nein: sein Leben von ihm selbst erzählt* [A little yes and a big no: His life told in his own words]. Hamburg: Rowohlt (Original work published 1955).

Hansen, M. (1999). Benjamin and cinema: Not a one-way street. *Critical Inquiry, 25*, 306–343.

Hansen, M. (2004). Room-for-play: Benjamin's gamble with cinema. *October, 109*, 3–45.

Hausmann, R. (1977). Dada empört sich, regt sich und stirbt in Berlin [Dada rebels, wells up, and dies in Berlin]. In K. Riha (Ed.), *Dada Berlin: Texte, Manifeste, Aktionen* (pp. 3–12). Stuttgart: Philipp Reclam.

Herzfelde, W. (1962). *John Heartfield: Leben und Werk* [John Heartfield: Life and work]. Dresden: Verlag der Kunst.

Hong, Y.-S. (1998). *Welfare, modernity, and the Weimar state, 1919–1933*. Princeton, NJ: Princeton University Press.

Huelsenbeck, R. (1987). Dadaistisches Manifest [Dadaist manifesto]. In R. Huelsenbeck (Ed.), *Dada Almanach: Poetische Aktion* (pp. 36–41). Hamburg: Edition Nautilus (Original work published 1920).

Jardine, N. (2000). *The scenes of inquiry: On the reality of questions in the sciences*. Oxford, England: Oxford University Press.

Kahn, E. (1919). Psychopathen als revolutionäre Führer [Psychopaths as revolutionary leaders]. *Zeitschrift für die gesamte Neurologie und Psychiatrie, 52*, 90–106.

Kaufmann, D. (1999). Science as cultural practice: Psychiatry in the First World War and Weimar Germany. *Journal of Contemporary History, 34*, 125–144.

Killen, A. (2003). From shock to Schreck: Psychiatrists, telephone operators and traumatic neurosis in Germany, 1900–1926. *Journal of Contemporary History, 38*, 201–220.

Knopf, J. (1974). *Bertolt Brecht—Ein kritischer Forschungsbericht. Fragwürdiges in der Brecht-Forschung* [Bertolt Brecht—A critical research report on questionable aspects in research on Brecht]. Frankfurt am Main: Fischer.

Kohler, R. (2002). *Landscapes and labscapes: Exploring the lab–field border in biology*. Chicago: University of Chicago Press.

Kracauer, S. (1998). *The salaried masses: Duty and distraction in Weimar Germany* (Q. Hoare, Trans.). London: Verso (Original work published 1930).

Lavin, M. (1993). *Cut with the kitchen knife: The Weimar photomontages of Hannah Höch.* New Haven, CN: Yale University Press.

Lerner, P. (1996). Rationalizing the therapeutic arsenal: German neuropsychiatry in World War I. In M. Berg & G. Cocks (Eds.), *Medicine and modernity: Public health and medical care in nineteenth- and twentieth-century Germany* (pp. 121–148). Cambridge, England: Cambridge University Press.

Lerner, P. (2000). Psychiatry and casualties of war in Germany, 1914–1918. *Journal of Contemporary History, 35,* 13–28.

Lerner, P. (2001). An economy of memory: Psychiatrists, veterans, and traumatic narratives in Weimar Germany. In A. Confino & P. Fritzsche (Eds.), *The work of memory: New directions in the study of German society and culture* (pp. 173–195). Urbana: University of Illinois Press.

Lerner, P. (2003). *Hysterical men: War, psychiatry, and the politics of trauma in Germany, 1890–1930.* Ithaca, NY: Cornell University Press.

Leslie, E. (2002). *Hollywood flatlands: Animation, critical theory, and the avant-garde.* London: Verso.

Matless, D., & Cameron, L. (2006). Experiment in landscape: The Norfolk excavations of Marietta Pallis. *Journal of Historical Geography, 32,* 96–126.

McCarren, F. (1998). *Dance pathologies: Performance, poetics, medicine.* Stanford, CA: Stanford University Press.

McCarren, F. (2003). *Dancing machines: Choreographies of the age of mechanical reproduction.* Stanford, CA: Stanford University Press.

Meskill, D. (2004). Characterological psychology and the German political economy in the Weimar Period (1919–1933). *History of Psychology, 7,* 3–19.

Micale, M. (2004). Introduction: The modernist mind—A map. In M. Micale (Ed.), *The mind of modernism* (pp. 1–19). Stanford, CA: Stanford University Press.

Möbius, H. (2000). *Montage und Collage: Literatur, bildende Künste, Film, Fotografie, Musik, Theater bis 1933* [Montage and Collage: Literature, fine arts, film, photography, music, and theater until 1933]. Paderborn, Germany: Wilhelm Fink Verlag.

Nolan, M. (1994). *Visions of modernity: American business and the modernization of Germany.* Oxford, England: Oxford University Press.

Ong, A., & Collier, S. (Eds.). (2004). *Global assemblages: Technology, politics, and ethics as anthropological problems.* London: Wiley.

Prinzhorn, H. (1922). *Bildnerei der Geisteskranken: Ein Beitrag zur Psychologie und Psychopathologie der Gestaltung* [Sculpture by the mentally ill: On the psychology and psychopathology of design]. Berlin: Springer.

Rabinbach, A. (1990). *The human motor: Energy, fatigue, and the origins of modernity.* New York: Basic Books.

Rajan, K. S. (2006). *Biocapital: The constitution of postgenomic life.* Durham, NC: Duke University Press.

Rheinberger, H. J. (1997). *Toward a history of epistemic things: Synthesizing proteins in the test tube.* Stanford, CA: Stanford University Press.

Roelcke, V. (2001). Electrified nerves, degenerated bodies: Medical discourses on neurasthenia in Germany, circa 1880–1914. *Clio Medica, 63,* 177–197.

Ronell, A. (2003a). The experimental disposition: Nietzsche's discovery of America (or why the present administration sees everything in the form of a test). *American Literary History, 15,* 560–574.

Ronell, A. (2003b). Proving grounds: On Nietzsche and the test drive. *Modern Language Notes, 118,* 653–669.

Ronell, A. (2005). *The test drive.* Urbana: University of Illinois Press.

Rose, N. (1999). *Governing the soul: The shaping of the private self* (2nd ed.). New York: Free Association Books.

Rose, N. (2006). *The politics of life itself: Biomedicine, power, and subjectivity in the twenty-first century*. Princeton, NJ: Princeton University Press.

Ryan, J. (1991). *Vanishing subject*. Chicago: University of Chicago Press.

Sadoff, D. (1998). *Sciences of the flesh: Representing body and subject in psychoanalysis*. Stanford, CA: Stanford University Press.

Santner, E. (1995). *My own private Germany: Daniel Paul Schreber's secret history of modernity*. Princeton, NJ: Princeton University Press.

Sass, L. (1992). *Madness and modernism: Insanity in the light of modern art, literature, and thought*. New York: Basic Books.

Schaffer, S. (1992). Self-evidence. *Critical Inquiry, 18*, 327–362.

Schaffer, S. (1997). Experimenters' techniques, dyers' hands, and the electric planetarium. *Isis, 88*, 456–483.

Schräge, D. (2001). *Psychotechnik und Radiophonie: Subjektkonstruktionen in artifiziellen Wirklichkeiten, 1918–1932* [Psychotechnics and telephony: Constructions of subjects in artificial realities, 1918–1932]. Munich: Fink.

Schwartz, F. (2000). The eye of the expert: Walter Benjamin and the avant-garde. *Art History, 24*, 401–444.

Shapin, S., & Schaffer, S. (1985). *Leviathan and the air-pump: Hobbes, Boyle, and the experimental life*. Princeton, NJ: Princeton University Press.

Sibum, O. (1995). Reworking the mechanical value of heat: Instruments of precision and gestures of accuracy in Early Victorian England. *Studies in the History and Philosophy of Science, 25*, 73–106.

Singer, K. (1919). Das Kriegsende und die Neurosenfrage [The end of war and the question of neuroses]. *Neurologisches Zentralblatt, 38*, 330–331.

Sloterdijk, P. (1987). *Critique of cynical reason* (M. Eldred, Trans.). Minneapolis: University of Minnesota Press.

Smith, C., & Agar, J. (1998). Introduction. In C. Smith & J. Agar (Eds.), *Making space for science: Territorial themes in the shaping of knowledge* (pp. 1–23). New York: St. Martin's Press.

Thiher, A. (1999). *Revels in madness: Insanity in medicine and literature*. Ann Arbor: University of Michigan Press.

Tucholsky, K. (1920, July 20). Dada. *Berliner Tageblatt*.

Vasudevan, A. (2006). Experimental Urbanisms: *Psychotechnik* in Weimar Berlin. *Environment and Planning D: Society and Space, 24*, 799–826.

Biesalski, K. (n.d.). *Krüppelsfürsorge und Sozialbiologie* [Care of the physically handicapped and social biology] (Pamphlet). Berlin: Oskar-Helene-Heim.

NGOs, the Science-Lay Dichotomy, and Hybrid Spaces of Environmental Knowledge

Sally Eden

In debates about science and the environment, the "science-lay dichotomy is both highly tenuous and highly tenacious" (Irwin & Michael, 2003, p. 124). It is tenacious because, despite continual criticism from social scientists, it continues to underpin the "cognitive-deficit model" of the public understanding of science. The deficit model rests on the assumption that the lay public is unscientific, unspecialized, and often ignorant (or at least poorly informed) about the details of scientific and technological developments and are therefore normally excluded from decisions about how science and the environment is managed. It is consequently also assumed in the model that this exclusion and lack of knowledge breed public distrust in scientific developments and their regulation and, therefore, that this distrust must be corrected by providing more information and improving public education about these matters.

The science-lay dichotomy is also tenuous because it considerably oversimplifies the complex reality of science–society relations. For example, it does not readily reflect the contribution made by environmental nongovernmental organizations (NGOs), which are key actors in the case of environmental science. The increasing recruitment of scientifically trained researchers into the large NGOs in particular and their deployment of scientific evidence in their campaigns make it difficult to see NGO personnel as strictly "lay." But it is often equally difficult to see them as strictly "scientific," for they rarely generate the science that they deploy, and they frequently challenge notions of expertise, scientific certainty, and issue closure in the interests of opening up environmental debates. Moreover, the self-education of local activists through their involvement with NGO activity also challenges the simplistic and monolithic construction of a supposedly "lay public" in environmental debates (see Gregory & Miller, 1998; Irwin & Michael, 2003).

S. Eden (✉)
Department of Geography, University of Hull, Hull HU6 7RX, UK
e-mail: s.e.eden@hull.ac.uk

P. Meusburger et al. (eds.), *Geographies of Science*, Knowledge and Space 3,
DOI 10.1007/978-90-481-8611-2_12, © Springer Science+Business Media B.V. 2010

I therefore consider how NGOs see and manage their place in relation to this science-lay dichotomy within what one might think of as the cultural cartographies of science (Gieryn, 1999) across multiple sociospatial contexts. To address the relative neglect of NGOs in the sociology of science literature (Jamison, 1996), I consider how far environmental NGOs can be considered to be scientific actors rather than, as is more common, political actors or social movements. I adopt Gieryn's (1983, 1995, 1999, 2008) notion of "boundary-work" to argue that NGOs hybridize the science-lay dichotomy in different ways. They not only move across that boundary frequently as translators or mediators (Latour, 2005)[1] but also seek hybrid spaces of science and lay knowledge to advance their own agendas for environmental reform, sometimes even deliberately creating new spaces for that purpose.

I reflect on these notions by drawing on two empirical studies of environmental NGOs that I undertook with colleagues. One study looks at general notions of science, expertise, and credibility among United Kingdom NGOs involved in waste debates in 2002–2003; the other, at environmental knowledge for certifying sustainable products through two specific international NGOs in 2004–2006. Although that research involved a great deal of qualitative data collection and analysis, I present little of it here. Instead I will briefly use the two studies to illustrate my conceptual argument.

NGOs, Environmental Science, and Boundary Work

I set this chapter within wider debates about the democratization of science (Beck, 1992, 1995; Funtowicz & Ravetz, 1993; Gibbons et al., 1994; Irwin, 1995; Irwin & Wynne, 1996; Jasanoff, 2003; Nowotny, 2003; Nowotny, Scott, & Gibbons, 2001). Many of these debates have been rehearsed elsewhere, so this chapter centers on the question of how NGOs might feature in such discussion about the science-lay dichotomy. More precisely, I inquire into how they can help challenge "the modernist tradition in distinguishing science, treated as a universal public good, from knowledge, which is seen by contrast as particularistic, indigenous or local" (Martello & Jasanoff, 2004, p. 336). Although "science" in its "pure" form can be highly exclusionary for actors such as NGOs (as I show below), "knowledge" is a more adaptable and inclusive term and thereby facilitates this challenge to modernism in the interests of democratizing environmental knowledge and the debates in which it figures.

Irwin and Michael (2003) criticize the dominance of the modernist science-lay dichotomization by focusing upon hybridity, heterogeneity, and the mixing of these two qualities within individuals and networks: "Instead of assuming the contrast between science and society, it becomes necessary to explore contrasts between actors or constituencies each comprised of mixtures of both science and society" (p. 111). Hence, these two authors direct attention instead to "coalitions or assemblages or nexûs" (p. 98), movements, intermediaries, and hybrid groups, which they refer to as "ethno-epistemic alliances" (p. 98) working across the lay-expert divide. These alliances deploy heterogeneous "understanding" that is not restricted to the cognitive understanding with which the deficit model is concerned but brings moral judgment and emotional response into play, much as Martello and Jasanoff (2004) emphasize knowledge rather than science.

This perspective relates well to NGOs. In the public domain, NGOs are frequently portrayed as antiscience or counterscience or as "emotional" because it is assumed that NGOs have developed as part of a backlash against scientism and particularly against the modernist approach of scientific and technocratic rationalization in environmental decision-making. But researchers have often been positive about how NGOs can contribute to geographies of science and environmental knowledge. Some observers propose that NGOs can open science up in a democratic challenge to help (or force) knowledge production to become more socially accountable and more diffuse than it usually is. Indeed, modern environmentalism flourished alongside similar "countercultural" challenges to the hegemony of science and technology in the 1960s (Eckersley, 1992; McCormick, 1995). Such a critique of science is also often coupled with the ability of NGOs to nurture and legitimate alternative spheres of knowledge. In this vein, Jamison (2001) argued that social movements such as environmentalism provide "a seedbed, or *alternative public space*, for the articulation of utopian 'knowledge interests'... [and] for the reconstitution of knowledge" (p. 46, italics added).

Jasanoff (1997) particularly considers the case of NGOs dealing in lay knowledge from indigenous people in developing countries. In this situation it is the boundary status of NGOs that enables these organizations to tap into knowledge outside the domain of pure science. Thus, "NGOs offer an important alternative to the standard top–down model of knowledge-making and knowledge diffusion in the natural sciences" and "constitute a vehicle for scaling knowledge up from the grass roots" (Jasanoff, 1997, p. 591). The knowledge with which NGOs deal may come from nonscientifically authorized sources (e.g., local communities) or may be of a different type (e.g., mixed knowledges like those meant by Irwin & Michael, 2003, which bring moral and ethical questions into debates dominated by scientific and technical questions). NGOs can thereby both broaden scientific knowledge and work to change science itself by hybridizing various spheres of knowledge and building the heterogeneity of debates and networks.

But such a challenge to the status quo can trigger a backlash, with critics denying that NGOs have the ability to produce science. That stance refutes the assertion that these organizations can understand, apply, and judge science and therefore positions them outside the sanctioned realm of scientific production. Characterizations of this sort are typical of boundary work (Gieryn, 1983, 1995, 1999; Jasanoff, 1987) and of the way zones (and representatives) of scientific authority and nonauthority are demarcated in order to enhance credibility. Critics like Turner (2001) have denounced claims that NGOs or other political actors such as the Sierra Club have expertise. He contends that their knowledge is intentionally oriented to policy rather than pure science and therefore "is at best part of the penumbral regions of scientific expertise" (p. 135). This criticism associates expertise with the classic Mertonian norm of disinterestedness (or interest in science for its own sake) and equates political utility with "contamination." Those individuals, groups, or organizations cast outside the boundary of expertise and authority must then work hard either to get back in or to realign the boundary more to their advantage. Such scientific boundaries are also contingent in time and space, are issue-specific, and are contingently validated and accepted in a mutual (but not necessarily friendly) co-construction of lay and expert actors (Michael, 2002; Gieryn, 1995).

Most research on boundary work has concentrated on how scientists deploy it in defense of science and how they institutionalize it through "boundary organisations" (Guston, 1999). Boundary organizations manage the interaction across the science–nonscience boundary by defining who can be involved how and by consolidating those rules of engagement through codes and networks. They therefore both permit and control transboundary exchange by providing "stable but flexible places in which scientists and other interested parties struggle over boundaries between science and politics," thus "capturing and routinizing the threatening heterogeneity of public claims at the science–politics boundary" (Kelly, 2003, p. 357). But, like more traditional scientific institutions (Kinchy & Kleinman, 2003; Moore, 1996), many boundary organizations preserve the modernist notion of science while simultaneously broadening the ways in which it is debated and promoting its usefulness to policy. And, of course, membership even of "hybrid" boundary organizations remains exclusive, often by invitation only.

By contrast, the position and increasing NGO recruitment of scientifically qualified staff in the 1990s and 2000s suggest that NGOs, because they are not seeking to stabilize the science-lay boundary, are far more heterogeneous than boundary organizations (as currently conceptualized), and operate differently than they do. Indeed, many environmental NGOs challenge boundary organizations, one example being Greenpeace's arguing with the United Kingdom's Royal Society over the scientific evidence indicating the potential environmental impact of genetically modified (GM) crops. NGOs may also be invited to participate in boundary organizations as nonscientific representatives but are less interested in the careful management of science than in the opportunities to challenge it and make it more useful for environmental issues.

Most of the boundary organizations that have been studied are scientifically based institutions, such as climate research groups (Agrawala, Broad, & Guston, 2001; Miller, 2001) and biomedical ethics committees (Kelly, 2003). Moreover, studies of boundary organizations, like much of the literature on boundary work, have concentrated upon how science, scientists, and scientific organizations communicate with the "outside" (e.g., Moore, 1996) rather than how the outside engages with science. This focus tends, perhaps unintentionally, to reinforce the science-lay dichotomy. In this chapter, however, I take the opposite view, considering how NGOs perceive and hybridize this boundary and the consequences for how they work. In particular, I examine what this perspective tells us about the place of NGOs in a cartography of credibility (Gieryn, 1999).

Being Outside Science

I start by considering the assertion that NGOs are increasingly seeking to recruit researchers with recognized scientific and professional training (e.g., Jamison, 2001; Yearley, 1993). For example, all but 4 of the 22 NGO interviewees in the 2002–2003 study (Eden, Donaldson, & Walker, 2006) had scientific qualifications, with first degrees in chemistry, physics, geology, ecology, environmental management, and similar subjects. Most of these people also had additional degrees (Masters, Ph.D., or both) in fields like environmental management, waste or environmental technology. Recruiting such people may be a form of boundary work, for in the eyes of a chemist holding a Ph.D. and working for a university research institute an NGO employee with a Ph.D. in waste technology may be more legitimate than someone without that background. (At least the recruiting NGO might hope so.) One can only speculate about this perception at the moment, for there are many accounts about scientists' socialization (e.g., Campbell, 2003) but few about scientists transferring their expertise to more heterogeneous networks such as NGOs.

Despite having scientific backgrounds, however, the interviewees from NGOs saw science as a group or sect that was closed to outsiders, including themselves. They felt themselves to be outside the boundaries of pure science:

> There's a whole culture of science which is [long pause], it's a different world, which you can feel excluded from. It produces a set of people who may not particularly well interact with people who don't know science. There's an awful barrier. If you haven't got a science background, you're terrified of being foolish and not understanding things. And so the way that it comes over to nonscientists is that it's hierarchical and autocratic. (Small environmental NGO in interview)

The interviewees regarded themselves as being outside science because they distinguished science as a process of generating original knowledge. What they did, by contrast, was not regarded as science; they saw it as merely secondary analysis. These NGO interviewees therefore accepted a very traditional view of modernist science and its authority, especially its independence. Hence, as separate from their own organizations "scientific expertise remains the principal form of legitimation in the leading environmental organisations... [T]here are no viable alternatives" (Yearley, 1991, p. 38).

Notably, some of the interviewed NGOs made much stronger scientific claims in public than in private. For example, WWF stated on its website that it was "the science-based conservation organisation" (retrieved April 2003 from http://www.wwf.org), but in the interview material that my research team and I obtained, interviewees identified their boundary condition as *outside* the purely scientific authority, as critics like Turner (2001) do.

But it is important to say that our interviewees also saw their outsideness positively because it gave them rhetorical room to question science where science was too narrow and restrictive for their purposes, especially in complex environmental policy debates. In this sense the NGOs seek a role comparable to that suggested above by Jasanoff (1997) and Jamison (2001)—that of providing political critique and an alternative space for knowledge generation and validation.

But entrenching scientific authority in environmental debates to the point that it is unquestioningly naturalized as the norm makes it difficult to validate alternatives. The media continue to portray environmental debates as arguments about scientific or technical disagreements, an approach that polarizes discourse (Smith, 2000). NGOs, to their own disadvantage, can therefore be drawn into arguments defined (and, therefore, restricted) by other boundary workers:

> I sometimes feel that the environment movement—and indeed ourselves—conspire to make environmental issues about scientific ones. Sometimes they're not. Sometimes they are about values, sort of moral treatment of the environment. And I think on GM we have actually ended up supporting the idea that somehow, if only the right science can be done, then we'd all know how it should be. But it's actually not so straightforward. (Large environmental NGO in interview).

The interviewees thus sometimes fought against restrictive scientific boundary work, but the struggle was not easy. Their situation differs from that of Epstein's (1995, p. 420) patient–activists, who did succeed in bringing moral and ethical arguments to bear on debates about the scientific validity of medical treatment trials. But Epstein's case, as he was at pains to point out, was unusual because the very condition and suffering of the patient–activists endowed them with legitimacy and warranted a claim to knowledge that scientific researchers could not have. By comparison, representatives of environmental NGOs are neither necessarily the ones suffering from pollution or environmental damage nor animals facing extinction, although they may campaign on behalf of both groups. The NGO representatives have no claims to knowledge that are superior to those of university or industrial researchers. In other words, gaining accountability and legitimacy for their environmental views can be a problem.

With the emergent interest in and protection of "indigenous knowledge" about the environment, being local sometimes addresses this problem of accountability and legitimacy for specific campaigns and grassroots activism, especially in the developing world (e.g., Agrawal, 2002; Jasanoff, 1997). Perhaps support for local expertise arises because classical scientific knowledge inevitably suffers from identifiable gaps in such contexts, providing opportunities for lay contributions (e.g., regarding species not yet recognized by taxonomy but potentially important for biodiversity or medicine). In the developed world indigenous knowledge is less likely to offer benefits that scientific knowledge has not already tapped. Indeed, in some developed contexts lay mobilization to defend local environments is more likely to be denigrated as NIMBYism[2] than valued as indigenous knowledge. The result is denial of a local environmental claim to legitimacy against the universalizing norms of modernist science. Including NGOs in a cartography of credibility is thus problematic and highly contingent, especially where critics cast them outside the boundary of science, knowledge, and authority.

Hybridizing and Translating

Instead of challenging their uneasy position in relation to scientific authority by seeking entry to the zone of pure science (the existence of which NGOs mainly accept), NGOs generally try to hybridize knowledge across the science-lay boundary and build alliances to facilitate that effort:

> We always want to, if you like, have scientists as allies, not as enemies. And the way we communicate with scientists is to try to show them that we use science in a rational and reasonable way, but we have a moral agenda also. (Large environmental NGO in interview)

This boundary work was prompted not least by the scientific community's inability to deal with particular environmental problems fully, given their transdisciplinary, highly complex nature: "There is therefore a trade-off in environmental studies, such as ecological footprinting, between being scientific and being "value-laden," at which point "science" becomes an "art"—an important but unclear line of demarcation" (large environmental NGO in interview).

For the most part, our interviewees saw themselves as partly scientific personnel sitting on the lay-expert boundary and engaging in bridging work. But by this very work as mediators, they complicate the assumption of that simplistic boundary despite sometimes using such a boundary when talking in interviews:

> Our role… is to simplify the science, to simplify the conclusions, to a lay level for politicians, journalists and public. (Large environmental NGO in interview)

> Interpreting your basic scientific evidence or research or whatever it happens to be in such a way that it is meaningful to the people that it is being addressed to [is difficult]. Sometimes I hear interviews by scientists where, even though I know what they're talking about, I can't understand a bloody word of it, you know. And that's *me*. I read scientific papers in the original. What's it like for 99.9% of other people who don't? (Large environmental NGO in interview)

One instance of this work was the FactoryWatch website (now closed), set up by Friends of the Earth and discussed by one interviewee. This group had built a database on industrial emissions, using data from the national regulatory body after having "cleaned" the input by working with that agency to clarify errors. Friends of the Earth had then put this resource on its website so that members of the general public could find the results for the areas in which they lived. Blurring the "lay-expert" divide, the project demonstrated "knowledge development as an active process in which more than simply scientists participate" (Irwin & Michael, 2003, p. 108). It also emphasized that the translational status of NGOs has merit: They can liaise with nonscientific groups about scientific information because their heterogeneous character encompasses both qualities, making them effectively bilingual.

> Sustainability and sustainable development are [regarded as] ugly terms[, but] they're not. They haven't got engagement with—certainly not with the general public. And even with relatively specialist audiences, people still struggle. So, when we're going in and talking

about sustainable development, that's quite new to nearly everyone we talk to. So as long as we are confident in our knowledge of sustainable development as a broad concept, we can help them and we can build knowledge and capacities within these companies without necessarily having to be specialists in the water sector or in the energy sector. (Small environmental NGO in interview)

This case shows that NGOs engage not simply in boundary work to move or reconfigure boundaries (as Collins & Evans, 2002, might suggest) but to create a hybrid space, a zone of heterogeneous knowledge practices. They seek to move beyond a modernist dichotomy to a fluid space in geographies of science, one predicated less on science than on potentially useful knowledge:

In communication to the public, that can... be very focussed along... lines [ranging from] "put this in that bin [for recycling]"... [and] "this is why" to communicating a whole waste strategy within which you've got information of a scientific nature, of a technological nature, of an economic nature and then almost communicating the recommendations. (Small not-for-profit waste group in interview)

NGOs work the boundary not merely to circulate factual information to the public but also to widen the circulation of information and ideas about policy, ethics, and practical application in order to mobilize other actors and publics, raise awareness, and encourage purposeful scientific input to key debates. Their translation is of "science in context," not merely science. In keeping with the preceding arguments, NGOs are distributing explicitly politicized information, whereas the deficit model rests on the false assumption that the provision of information is apolitical. The process is not one of just passing scientific information on to various publics but of *translating* it (a function that implies both movement and change) so that it relates to environmental agendas and becomes useful. As Jasanoff (1997) notes, it is "this link between knowledge and action that provides environmental NGOs their primary point of political intervention" (p. 580).

This intervention relies heavily upon popularization by which information is translated between scientific and public arenas. According to Hilgartner (1990; also Gregory & Miller, 1998), the popularization of science is criticised in the deficit model of public understanding because scientists regard products of popularization (whether on TV, in the press, or on the Internet) as inferior to their own (e.g., academic articles and scientific publications). The further downstream the translations go toward the public, the lower they rate. But Hilgartner argues that popularization is practiced both within science and between science and its publics. Even specialized articles in scientific journals summarize, simplify, and omit information (and judgments) for the sake of clearly communicating with and convincing other scientists. When scientists cite other scientific articles in their own work, these downstream citations are necessarily simplified—often to corroborate their own arguments—so "'popularisation' is simply a matter of degree" (Hilgartner, 1990, p. 528). It is even more evident in interdisciplinary fields when scientists communicate to other scientists outside their specialty in the service of the wider goal or policy. As Gregory and Miller (1998) pointedly note, "when it comes to biology, physicists are laypeople" (p. 97); no one is a specialist in every scientific field. Hence, popularization *is not separate from* producing scientific knowledge, for "scientists who popularize are doing science in public" (p. 84, italics added).

This notion applies rather well to NGOs in the sense that actively constructing and challenging science through public debate and popular communication can be seen as participating in science rather than merely criticizing or corrupting it.

> We're spending a lot of time saying to people like DEFRA [Department for the Environment, Food and Rural Affairs, a government ministry], Treasury, DTI [Department for Trade and Industry, a government ministry at the time], we're saying to them, "No, the data says. . ." We're saying, "No, interpret it this way; here's some new data that we've commissioned ourselves and debate." But our position in all of this is saying, rather than the people who present science and say, "There it is." [We're] the people who actually say, "What does that mean in terms of the real world for policy-making and for decisions over whether firms should or shouldn't be allowed to commercialise GM or have dioxins coming out of the top of incinerator stacks?" or whatever it happens to be. That's where we're at really. We're in that public, that interpretation of it, in a sense. (Large environmental NGO in interview).

Moreover, NGOs must practice "appropriate simplification": Knowledge must be distilled to be succinct and usable by popular audiences (Hilgartner, 1990, p. 519) rather than left in the esoteric circumstances of its production. The details of academic science and research on, say, the mobility of heavy metal ions in incinerator fly ash or life-cycle analysis measurements are "of completely no interest at all to most of the people that we normally deal with; it's so obscure" (small waste NGO in interview) or impenetrable because of "such a high fog factor" (waste association in interview). Without translation, science, however rigorous and accurate, is unusable: "I don't see the point of science if you can't communicate it properly" (small environmental NGO in interview).

> [My question is] how do we use science to communicate to us as individuals that need to change behavior, rather than saying "an incinerator produces x milligrams of dioxin per meter cubed?" So my approach generally is to use science in a way that will encourage the change of behavior. (Small waste NGO in interview)

But NGOs exercise "pragmatic epistemological flexibility" (Yearley, 1996, p. 183) in using or contesting science, depending upon the issue being debated. In other words, the positioning of each NGO in relation to science can vary. For example, the current furore in the United Kingdom over wind farms divides the conservation lobby into those people and groups supporting wind farms in the interests of climate change and energy conservation and those opposing wind farms in the interests of landscape aesthetics and bird protection. Boundary work reflects and supports not only the definition of science but also differentiation among NGOs, which is itself seen positively: "It's been my view, having seen. . . the 10 years of the spectrum of NGOs, that the deep greens are needed because they are bringing the [very] light. . . green to a darker level, because we don't look so radical [in comparison]" (waste NGO in interview).

Creating New Hybrid Spaces

I have suggested that NGOs, rather than seeking inclusion within the boundary of science, endeavor to hybridize the science-lay dichotomy for their own purposes (with variable success depending on the issue). In some cases, however, NGOs go further and deliberately create new hybrid spaces of environmental knowledge practices for specific uses. For example, a coalition of NGOs, including some of those NGOs quoted above, supports the Forest Stewardship Council (FSC), which administers a certification system for products from sustainably managed forests, and the Marine Stewardship Council (MSC), which runs an analogous system for products from sustainably managed marine fisheries.

Without going into the detail of their systems, I use the FSC and the MSC to illustrate my general argument briefly in terms of how NGOs can bring knowledge to bear on an environmental issue.[3] Both the FSC and the MSC are global networks. As such, they are not bound by territorial sovereignty; they set standards for sustainable management across a wide range of different environments, political regimes, and sociotechnical systems. They do so by negotiating multiple levels of influence and developing new mechanisms and new "spheres of authority" (Pattberg, 2005, p. 177) that are credible. To exert influence, both the FSC and the MSC must build such authority without a government's threat of coercion, relying instead on persuasion and moral legitimacy, especially by building "transparency, reliability, and neutrality" (Pattberg, 2005, p. 181).

Part of this task involves invoking science as an important source of credibility, for it has associated norms of authority, expertise, disinterestedness, and objectivity. For example, the involvement of science and scientists at every level and function of the MSC is underscored throughout this organization's literature to support the claim that its standard is based on "an objective and scientifically verifiable method of assessing the sustainability of fisheries" and that the Council draws on "an international panel of experts from the fisheries, environmental and governmental scientific communities, together with representatives from the catching, processing and retailing sectors" (Marine Stewardship Council, 2006, p. 2). Both NGOs convene "expert panels" of various sorts, particularly to review controversial issues (e.g., Forest Stewardship Council, 2007). These NGOs thus strive for that mixture of science and society suggested by Irwin and Michael (2003).

The standards of the FSC and MSC vary in time and space to keep pace with changes in scientific, technological, and environmental knowledge and practice. FSC National Standards are required to be updated regularly, and both the FSC and the MSC undergo recertification every 5 years. Particularly important issues that these organizations monitor are the acceptable use and types of pesticides, the amount of deadwood needed per hectare to support biodiversity, the nesting territories of rare or endangered birds, and the level and potential for regeneration of fish stocks in the oceans.

The FSC and the MSC have adopted third-party certification as the gold standard of credibility (Cashore, Auld, & Newson, 2004; Hatanaka, Bain, & Busch, 2005; Jahn, Schramm, & Spiller, 2005). Certified forests and fisheries (and products from them) are checked by a body independent of the manufacturer or retailer. This body functions as a quasi-regulatory agency or police force that verifies whether applicants meet the standards. These NGOs are not merely lobbyists or researchers but also appliers of environmental principles and auditors and managers of verification systems. This approach, too, calls attention to the heterogeneity of knowledge practices and personnel within this diverse sector.

Such activities do not reach out to different publics just to disseminate environmental knowledge, as the deficit model might indicate. Rather, they bring different publics inside a new space, a new institution, and a new set of verification processes in order to translate and transform (Latour, 2005) environmental knowledge into specific practices. The FSC and MSC become hubs of hybrid knowledge, clearing houses for diverse sources ranging from practical local knowledge about how trees grow in California to reviews of scientific journals about the ideal amount of deadwood in a forest to support biodiversity. But the boundaries of these hybrid spaces are less well defined than others. They do not invoke modernist dichotomies such as lay-expert, science–nonscience, or science–politics. Instead, they are fuzzy, highly dynamic, adaptable, and, unlike the general NGO discussion above, intimately connected to practice. The knowledge they demarcate is produced in and for context (see Nowotny et al., 2001), but it is a context of forests and oceans that belong to a network. These boundaries may be beginning to define alternative spaces for reconstituting knowledge, as Jamison (2001) suggests, although the way in which that membership is managed makes these spaces far from fully public despite their obvious heterogeneity.

Conclusion

Jasanoff (1997) argues that the ability of environmental NGOs to "bridge the lay-expert, activist–professional and local–global divides" (p. 581) can make epistemic groups more inclusive than they are. But NGOs do not merely bridge these dichotomies; they hybridize them, sometimes developing new networks and spaces in which to increase their success through a "concrescence" of different kinds of knowledge (Michael, 2002, p. 372).

Concrescence can be important in multi-issue NGOs, like WWF, Greenpeace, and Friends of the Earth, whose knowledge must be wide enough to respond to the diverse and rapidly changing topics in public debates. It can be important also in issue-specific NGOs, like the FSC and the MSC, whose knowledge centers chiefly on particular environmental problems. NGOs of that type seek to reach out to hybridize and translate environmental debates in other, especially public spaces, such as the media and government policy circles. Unusually for issue-specific NGOs, the FSC and the MSC internalize that hybridity within their governance and knowledge-management structures. They often do so in quite closed spaces that are defined by membership and therefore more exclusive than is preferred by some commentators writing about the democratization of science. (The framework of these two NGOs is still more heterogeneous than what preceded them, however.) Such differences between NGOs also show that NGOs work in particular space-time contexts, rendering all attendant analysis "contingent, situated and reflexive" (Irwin & Michael, 2003, p. 154).

Emphasizing hybridity and Irwin and Michael's (2003) mixed alliances can help researchers go beyond the lay-expert dichotomy and its modernist problems, particularly when one theoretically positions NGOs as *both* lay and expert. Moreover, the hybrid spaces that such processes create and influence are highly contingent on continual investment and reinvestment in their survival and exposed to harsh external criticism. Such spaces are alternative knowledge spaces (Jamison, 2001), but they are not necessarily public ones, created as they are by invitation and subject to clear lines of demarcation and governance.

I therefore do not want to suggest that such hybrid spaces are a simple panacea to the modernist assumption of a problematic science-lay dichotomy. They are in many ways very difficult to build and sustain. My point is that recognizing and pursuing hybridity in these rather fluid ways may be helpful in challenging outdated models of the place science has in society and of the ways in which cultural cartographies of science are shaped and invoked.

Acknowledgment This paper comes from work funded by the United Kingdom's Economic and Social Research Council through its Science in Society Programme, awards L144250047 and RES-151-25-00035. I am grateful to all the interviewees for their time and to Andrew Donaldson, Christopher Bear, and Gordon Walker for their input to these projects over the years.

Notes

1. Latour (2005) used the word "intermediaries" (p. 39 passim) specifically to stand for entities that pass on meaning to others but without transforming it and reserved the term "mediators" (p. 39 passim) to refer to entities that transform meaning as they convey it to others. Irwin and Michael (2003) do not make this distinction, but it is likely that their intermediaries are, in effect, Latour's mediators, for few coalitions remain unchanged in the course of their work. In the rest of this chapter, I use *mediators* to avoid confusion among readers following Latour (2005).

2. The acronym stands for "not in my backyard" and is often used to undermine objections by local residents to projects believed to be dangerous, unsightly, or otherwise undesirable (e.g., incinerators, prisons, or homeless shelters). Its implication is that objections are prompted by self-interest rather than general principles.

3. For more details on these two examples, see particularly Eden (2009), and Bear and Eden (2008).

References

Agrawal, A. (2002). Indigenous knowledge and the politics of classification. *International Social Science Journal, 54*, 287–297.

Agrawala, S., Broad, K., & Guston, D. H. (2001). Integrating climate forecaster and societal decision making: Challenges to an emergent boundary organization. *Science, Technology, and Human Values, 26*, 454–477.

Bear, C., & Eden, S. (2008). Making space for fish: The regional, network and fluid spaces of fisheries certification. *Social and Cultural Geography, 9*, 487–504.

Beck, U. (1992). *Risk society*. London: Sage.

Beck, U. (1995). *Ecological politics in an age of risk*. London: Polity.

Campbell, R. A. (2003). Preparing the next generation of scientists. *Social Studies of Science, 33,* 897–927.

Cashore, B., Auld, G., & Newson, D. (2004). Legitimizing political consumerism: The case of forest certification in North America and Europe. In M. Micheletti, A. Follesdal, & D. Stolle (Eds.), *Politics, products, and markets: Exploring political consumerism past and present* (pp. 181–199). New Brunswick, NJ: Transactions Publishers.

Collins, H. M., & Evans, R. (2002). The third wave of science studies: Studies of expertise and experience. *Social Studies of Science, 32,* 235–296.

Eckersley, R. (1992). *Environmentalism and political theory*. London: UCL Press.

Eden, S. (2009). The work of environmental governance networks: The case of certification by the Forest Stewardship Council. *Geoforum, 40,* 383–394.

Eden, S., Donaldson, A., & Walker, G. (2006). Green groups and grey areas: Scientific boundary work, nongovernmental organisations, and environmental knowledge. *Environment and Planning A, 38,* 1061–1076.

Epstein, S. (1995). The construction of lay expertise: AIDS activism and the forging of credibility in the reform of clinical trials. *Science, Technology, and Human Values, 20,* 408–437.

Forest Stewardship Council. (2007). *Pesticides review*. Bonn: Forest Stewardship Council.

Funtowicz, S. O., & Ravetz, J. R. (1993). Science for the post-normal age. *Futures, 25,* 739–755.

Gibbons, M., Limoges, C., Nowotny, H., Schwartzman, S., Scott, P., & Trow, M. (1994). *The new production of knowledge*. London: Sage.

Gieryn, T. F. (1983). Boundary-work and the demarcation of science from non-science: Strains and interests in professional ideologies of scientists. *American Sociological Review, 48,* 781–795.

Gieryn, T. F. (1995). Boundaries of science. In S. Jasanoff, G. E. Petersen Markle, J. C. Petersen, & T. Pinch (Eds.), *Handbook of science and technology studies* (pp. 393–444). London: Sage.

Gieryn, T. F. (1999). *Cultural boundaries of science: Credibility on the line*. Chicago: University of Chicago Press.

Gieryn, T. F. (2008). Cultural boundaries: Settled and unsettled. In P. Meusburger (Series Eds.) & P. Meusburger, M. Welker, & E. Wunder (Vol. Eds.), *Knowledge and space: Vol. 1. Clashes of knowledge: Orthodoxies and heterodoxies in science and religion* (pp. 91–99). Dordrecht: Springer.

Gregory, J., & Miller, S. (1998). *Science in public: Communication, culture, and credibility*. New York: Plenum Trade.

Guston, D. H. (1999). Stabilizing the boundary between US politics and science: The role of the Office of Technology Transfer as a boundary organization. *Social Studies of Science, 29,* 87–111.

Hatanaka, M., Bain, C., & Busch, L. (2005). Third-party certification in the global agrifood system. *Food Policy, 30,* 354–369.

Hilgartner, S. (1990). The dominant view of popularization: Conceptual problems, political uses. *Social Studies of Science, 20,* 519–539.

Irwin, A. (1995). *Citizen science*. London: Routledge.

Irwin, A., & Michael, M. (2003). *Science, social theory and public knowledge*. Maidenhead, England: Open University Press.

Irwin, A., & Wynne, B. (Eds.). (1996). *Misunderstanding science? The public reconstruction of science and technology*. Cambridge, England: Cambridge University Press.

Jahn, G., Schramm, M., & Spiller, A. (2005). The reliability of certification: Quality labels as a consumer policy tool. *Journal of Consumer Policy, 1325,* 53–73.

Jamison, A. (1996). The shaping of the global environmental agenda: The role of nongovernmental organisations. In S. Lash, B. Szerszynski, & B. Wynne (Eds.), *Risk, environment and modernity* (pp. 224–245). London: Sage.

Jamison, A. (2001). *The making of green knowledge*. Cambridge, England: Cambridge University Press.

Jasanoff, S. (1987). Contested boundaries in policy-relevant science. *Social Studies of Science, 17*, 195–230.

Jasanoff, S. (1997). NGOs and the environment: From knowledge to action. *Third World Quarterly, 18*, 579–594.

Jasanoff, S. (2003). Breaking the waves in science studies: Comment on HM Collins and Robert Evans, 'The third wave of science studies'. *Social Studies of Science, 33*, 389–400.

Kelly, S. E. (2003). Public bioethics and publics: Consensus, boundaries, and participation in biomedical science policy. *Science, Technology, and Human Values, 28*, 339–364.

Kinchy, A. J., & Kleinman, D. L. (2003). Organizing credibility: Discursive and organizational orthodoxy on the borders of ecology and politics. *Social Studies of Science, 33*, 869–896.

Latour, B. (2005). *Reassembling the social: An introduction to actor-network-theory.* Oxford, England: Oxford University Press.

Marine Stewardship Council. (2006). *Managing fisheries for the future with the MSC.* London: Marine Stewardship Council.

Martello, M., & Jasanoff, S. (2004). Conclusion: Knowledge and governance. In M. Martello & S. Jasanoff (Eds.), *Earthly politics: Local and global in environmental governance* (pp. 335–349). Cambridge, MA: MIT Press.

McCormick, J. (1995). *The global environmental movement* (2nd ed.). Chichester, England: Wiley.

Michael, M. (2002). Comprehension, apprehension, prehension: Heterogeneity and the public understanding of science. *Science, Technology, and Human Values, 27*, 357–378.

Miller, C. (2001). Hybrid management: Boundary organizations, science policy, and environmental governance in the climate regime. *Science, Technology, and Human Values, 26*, 478–500.

Moore, K. (1996). Organizing integrity: American science and the creation of public interest organizations, 1955–1975. *American Journal of Sociology, 101*, 1592–1627.

Nowotny, H. (2003). Democratising expertise and socially robust knowledge. *Science and Public Policy, 30*, 151–156.

Nowotny, H., Scott, P., & Gibbons, M. (2001). *Rethinking science: Knowledge and the public in an age of uncertainty.* Cambridge, England: Polity Press.

Pattberg, P. (2005). What role for private rule-making in global environmental governance? Analysing the Forest Stewardship Council (FSC). *International Environmental Agreements, 5*, 175–189.

Smith, J. (Ed.). (2000). *The Daily Globe.* London: Earthscan.

Turner, S. (2001). What is the problem with experts? *Social Studies of Science, 31*, 123–149.

Yearley, S. (1991). Greens and science: A doomed affair? *New Scientist, 131* (1777), 37–40.

Yearley, S. (1993). Standing in for nature: The practicalities of environmental organizations' use of science. In K. Milton (Ed.), *Environmentalism: The view from anthropology* (pp. 59–72). London: Routledge.

Yearley, S. (1996). Nature's advocates: Putting science to work in environmental organisations. In A. Irwin & B. Wynne (Eds.), *Misunderstanding science? The public reconstruction of science and technology* (pp. 172–190). Cambridge, England: Cambridge University Press.

Regulatory Science and Risk Assessment in Indian Country: Taking Tribal Publics into Account

Ryan Holifield

Studies of the geography of scientific knowledge production have shown how securing the credibility and objectivity of a scientific claim requires erasing or masking traces of the "local." In order for a claim to be credible and objective, it needs to be true everywhere, not just in the place it was formulated (Latour, 1987; Law & Mol, 2001; Livingstone, 2003). Otherwise, the claim is doomed to remain "local knowledge": subjective, place-bound, and unverifiable. Conventional views of science suppose that true claims—by virtue of their production by trained, impartial observers applying the correct methods—simply transcend the particularities of a local context of discovery. By contrast, social, cultural, and geographical studies of science have demonstrated that delocalizing claims requires considerable work, which includes the creation and circulation of standardized equipment and what Latour (1987, 1999) dubbed immutable mobiles (see also Bowker & Star, 1996; Law & Mol, 2001; Livingstone, 2003; Rouse, 1987).

In this chapter I contend that a rather different set of geographic requirements for credibility and legitimacy emerge in the domain of what Jasanoff (1990, 1995) calls *regulatory science*. Regulatory science, which in the literature on geographies of science has received considerably less attention than academic science, refers to scientific activities conducted in order to provide a basis for public policy decisions. It "routinely operates with different goals and priorities and under different institutional and temporal constraints from science done in academic settings and without implications for policy" (Jasanoff, 1995, p. 279). For example, while academic science depends primarily on peer review to evaluate quality, regulatory science involves other procedures that extend beyond the community of scientists, such as agency guidelines and protocols, audits, judicial review, legal tests of adequacy, and legislative oversight. Because it is produced for the purpose of complying with statutory requirements, regulatory science is subject to stricter deadlines and more intense political pressure and is "particularly susceptible to divergent, socially conditioned interpretations" (Jasanoff, 1995, p. 282). In general, public accountability

R. Holifield (✉)
Department of Geography, University of Wisconsin-Milwaukee, Milwaukee, WI 53201-0413, USA
e-mail: holifiel@uwm.edu

P. Meusburger et al. (eds.), *Geographies of Science*, Knowledge and Space 3, DOI 10.1007/978-90-481-8611-2_13, © Springer Science+Business Media B.V. 2010

plays a more significant role in securing the credibility and legitimacy of methods and claims in regulatory science than it does in academic science. On the one hand, the legitimacy of regulatory science in part depends as much as academic science does on making techniques and standards universal, although in regulatory science "universal" is typically limited to the territory under a nation-state's regulatory jurisdiction. On the other hand, the credibility of claims in regulatory science also relies on appropriately localizing the same techniques and standards, that is, making adjustments that properly take into account local heterogeneity and difference.

One of the classic exemplars of regulatory science—and an important site of debate on how to adjust techniques and standards to local differences—is human health risk assessment (Jasanoff, 1986). Within the US Environmental Protection Agency (EPA), the purpose of risk assessments is to estimate the likelihood of harm from exposure to toxic agents present in the environment. In recent efforts to reform the risk assessment process, EPA has faced contradictory pressures. Regulated industries have demanded that EPA standardize the models and assumptions it uses in site-specific risk assessments, whereas many communities with contaminated sites have insisted that such models and assumptions be customized to place-specific particularities. This chapter examines how the debate over how to localize the risk assessment process has unfolded in the unique context of Indian Country[1] in the United States. I argue that in this context the key requirement for EPA to secure the credibility of human health risk assessment has been to localize it in such a way that it can take distinctive *tribal publics* into account.

The localization of human health risk assessment in Indian Country has two dimensions. First, tribal representatives have called on EPA to customize its approach to risk assessment to particular places—especially reservations—in order to assess and mitigate the risks to a set of publics quite different from the "general public." That set comprises members of various tribes practicing what have come to be called *tribal traditional lifeways*. Second, tribal representatives have emphasized that in Indian Country EPA must also localize public involvement in decision-making about risk assessment policy and practice in distinctive ways. One of them is that of convening dialogue at the right locations and scales and with appropriate spatial arrangements. In order to conceptualize these two dimensions of the localization of regulatory science, I build on aspects of Bruno Latour's (2004) conception of the *collective*, which I discuss further below.

Localizing Latour's Collective

The idea that science and scientists are simply neutral inputs into public policy-making, transparently representing the facts of nature in debates over societal values, still has many adherents. Over the past few decades, however, this conventional model of the relationship between science and the public has been challenged by a vast literature in science studies and related fields (e.g., Jasanoff, 2004). My intention is not to review this literature in this chapter but instead to explore the theoretical possibilities of one contribution to the debate over "how to bring the sciences into democracy": that of Bruno Latour (2004).

Latour's (2004) focus is the "modernist constitution" (p. 54) that separates the world into nonhuman nature and its facts (the domain of science) on the one hand and human society and its values (the domain of democracy) on the other (p. 54). In place of this ontological division, he offers the concept of the collective: "the work of collecting into a whole" (p. 59). This work, which Latour also describes as "progressive composition of the common world" (p. 59), consists not of the appeal to facts to settle disputes about values but rather of two powers: "taking into account" and "arranging in rank order" (pp. 102–116, especially p. 115). Both powers, in turn, have two requirements, each of which incorporates aspects of both fact and value. The first requirement of taking into account is what Latour calls *perplexity*, or the insistence that deliberation over the common world must not prematurely close the question of what constitutes reality. Perplexity reflects the constant emergence of new entities as candidates for membership in the "external reality" that later comes to be taken for granted: germs and viruses, genes and prions, planets and black holes, missing links, and previously undiscovered species, for example. The second requirement of taking into account is *consultation*, or the insistence that deliberation over the common world must not prematurely limit the number of voices that will participate in the work of collecting. The second power, that of arranging in rank order, consists of *hierarchization* (determining what is compatible with the common world) and *institution* (bringing closure to the question of what legitimately composes the common world).

My focus in this chapter is the first of the two powers (taking into account), as articulated by EPA's Tribal Science Council: the requirement to take tribal publics into account. The Council's requirement to localize the practice of risk assessment to make it sensitive to locally distinctive human-environment relationships—specifically, to tribal traditional lifeways—corresponds to Latour's requirement of perplexity. In that context, the question of how to localize is that of how to place and scale the public whose health risk will be assessed: the public as what researchers conventionally call an "object" of knowledge. By contrast, the Council's requirement to localize risk assessment decision-making to include the voices of locally distinctive polities or publics—specifically, tribes as "nations within" (Deloria & Lytle, 1998)—corresponds to the requirement of consultation. In this regard, the issue of how to localize is that of how to place, locate, and scale the public that participates in the development of risk assessment policies and practices: the public as "subject" of knowledge.

Risk Assessment, Tribal Traditional Lifeways, and the Publics of Indian Country

Risk Assessment and Public Engagement in the US Environmental Protection Agency

EPA pioneered the use of risk assessment and instituted it as the privileged basis for decision-making in its programs during the 1980s. EPA conducts two kinds of risk assessment: human health risk assessment, which focuses primarily on the likelihood of contracting cancer, and ecological risk assessment, which analyzes risks of damage to ecosystems. It conducts risk assessments both for individual chemicals, in order to set national standards for "acceptable levels," and for contaminated sites, which may involve multiple chemicals. Because EPA's approach to risk assessment relies on numerous policy-based assumptions designed to be protective, it is typically characterized as a hybrid of science and policy. Consequently, its risk assessment policies and practices have drawn fervent critique from regulated industries, which argue that they are overly protective, and from environmental activists, who contend that they are not protective enough (Kuehn, 1996).

Under the Clinton administration during the 1990s, EPA responded to these critiques by initiating reforms in its approach to risk assessment, particularly within the federal Superfund program established to clean up hazardous waste sites. Among the most important of these reforms was one designed both to standardize risk assessments nationwide and to promote community involvement in developing site-specific risk assessments.[2] Given the distinctiveness of the US political and regulatory system, under which regulatory officials must routinely make controversial decisions in "public, adversarial forums" (Jasanoff, 1986, p. 30), EPA's reform undoubtedly addressed the aim of bolstering the credibility and legitimacy of risk assessments in the face of growing public skepticism (Jasanoff, 1986; Wynne, 1987). On the one hand, standardization responded to the critique from industry that risk assessments were inconsistent and lacking in scientific rigor; on the other hand, promoting community involvement responded to activists' contention that accountable risk assessment requires democratic participation from stakeholders.

Self-Governance and Environmental Regulation in Indian Country

While EPA's institutionalization of risk assessment was proceeding, significant changes in the territorialization of Indian Country were also taking place. American Indian reservations and tribal lands in the United States are distinctive because they are settings in which many tribal members continue to practice tribal traditional lifeways. But they are also distinctive because they have been established by treaties with the US government as legal and political territories quite separate from the states, as "nations within" (Deloria & Lytle, 1998). These treaties recognize both the inherent sovereignty of tribes within their own territories and the legal rights of members to practice traditional lifeways on tribal lands. Because of the unique legal and political status of tribes, tribal members are simultaneously part of the general public of US citizens and the specific publics of tribal nations (Biolsi, 2005; Deloria & Lytle, 1998). In the latter sense, tribal governments do not represent subdivisions of the general public but distinctive polities separate from the general public. Consequently, tribes emphasize that EPA, in order to secure credibility and legitimacy in Indian Country, must guarantee not only public participation in regulatory science but also appropriate government-to-government consultation.

The territorialization of tribal land in the United States has a long and complex history, which I can only briefly summarize here (see Deloria & Lytle, 1998; Frantz, 1999; Silvern, 1999, among others). Most reservations were established and defined in the mid-nineteenth century by treaties between tribes and the US government. These treaties drew legal boundaries marking the limits of territorial spaces in which tribes—now defined as domestic dependent nations within US territory—could remain culturally and politically separate from the rest of the country, preserving their own practices of governing and interacting with the environment. However, the significance of reservation boundaries changed dramatically between the 1880s and the 1930s, as new federal laws undermined not only tribal separateness and autonomy but also tribes' landholdings, languages, and traditional ways of life. Unlike the Indian Reorganization Act of 1934, which restored a degree of self-determination to reservations, the US government's policies of the 1950s and early 1960s were designed to terminate the relationship between the federal government and Indian tribes by promoting the assimilation of tribal members and eliminating the reservation as a distinctive, autonomous space.

The late 1960s ushered in the "self-determination era" for American Indian reservations, and in this context tribes have asserted both their sovereignty within reservation territory and their treaty-protected rights to practice traditional ways of life on tribal lands. After suffering drastic budget cuts under the two Reagan administrations during the 1980s, tribes began to receive renewed federal support for self-governance beginning in the late 1980s and early 1990s. The Tribal Self-Governance Demonstration Project Act of 1991, for instance, established new funding and programmatic support for tribal governments to develop greater autonomy. Many reservations also benefited from the Indian Gaming Regulatory Act of 1988, which recognized the right of reservations to host casinos as new sources of revenue.

By the early 1990s, environmental contamination on tribal land had become a matter of grave concern within Indian Country. The new sources of support for tribal governments enabled many tribes to hire environmental protection specialists and become more involved in regulatory programs administered by EPA. They also enabled tribes to develop intertribal networks dedicated to environmental protection. In 1991, for instance, seven tribes established the National Tribal Environmental Council (NTEC), a membership organization to support the environmental protection and preservation efforts of individual tribes. Three years later, the tribes and EPA created the National Tribal Operations Committee to coordinate communication on regulatory programs between tribal agencies and the federal government.

Perplexity: Localizing Regulatory Science to Assess Tribal Publics at Risk

During the 1990s, tribes began to contend that EPA's scientific activities might be inappropriate or inadequate in the context of tribal land. Three major issues became particularly important. First, tribes expressed concerns about the agency's collection of scientific data that might be confidential or culturally inappropriate. Second, they asserted the importance of traditional knowledge as a valid source of environmental data. Third, they questioned the capacity of EPA's models and assessment tools to integrate the distinctive aspects of tribal traditional lifeways. Tribal traditional lifeways encompass not only practices of hunting, fishing, and gathering but also worldviews that refuse to separate human communities from the nonhuman environments that surround and sustain them (National EPA-Tribal Science Council, 2006b).

One focus of critiques by tribes was the use of standardized exposure scenarios in EPA's risk assessments. Risk assessments use exposure scenarios to calculate the amount of a toxic substance to which an individual is likely to be exposed over a given period of time. Site-specific risk assessments, for instance, might include exposure scenarios for a range of subpopulations engaging in different activities: on-site workers, adult residents living near the site, children living and playing near the site, and so forth. In most cases, instead of basing these exposure scenarios on site-specific data, EPA uses a set of standard default exposure factors meant to be applicable throughout the country. EPA's *Exposure Factors Handbook* (1997) includes standard parameters for body weight, life expectancy, inhalation rate, rate of soil ingestion, rate of fish consumption, and numerous other indicators of exposure to chemicals in the environment.

Prompted by concerns about the applicability of such parameters in the context of tribal land, risk assessors working with and for tribes began to develop what they called Native American subsistence scenarios or tribal traditional lifeways scenarios (Harper, Flett, Harris, Abeyta, & Kirschner, 2002; Harris, 2000; Harris & Harper, 1997). A key premise of tribal traditional lifeways exposure scenarios is that the standard default residential exposure factors compiled in the *Exposure Factors Handbook* are based on distinctively suburban patterns of human-environment interaction. Harris and Harper (1997) contend that tribal members practicing traditional lifeways experience higher exposures than the general public—including other American Indians—living in off-reservation suburban settings. First, traditional tribal members come into contact with contamination through multiple environmental pathways (e.g., gathered plants, wild game, and sweat lodges) that would play no role in a suburban residential exposure scenario. In addition, because traditional tribal members engage more frequently in outdoor cultural and work activities than typical suburban populations do, they ingest more water and soil and inhale higher quantities of air.

A second premise of traditional tribal exposure scenarios is that EPA's conventional approach to exposure assessment, with its emphasis on exposure to individuals and to populations, fails to recognize risks to the ties that bind people and environments together at the community scale. The assumption underlying the EPA's exposure assessments is that protecting individuals and subpopulations likely to face the highest exposures protects the entire affected community.

In contrast, the developers of tribal traditional exposure scenarios conceptualize tribal communities as webs of relations, encompassing humans and nonhumans. They emphasize that exposure to a particular individual in a tribal community, such as an elder who alone possesses particular knowledge of the local environment, might place the entire web of relations at risk. In addition, they contend that because "human beings cannot be truly separated from the environment, it is inherently unsatisfactory to focus on human exposure as isolated from environmental effects" (Harris & Harper, 1999, p. 2). Although the developers of these scenarios do not claim that all risks to such webs of relations—or "cultural risks"—can be quantified, they suggest that risk assessors must somehow find ways to translate such relations into exposure assessments. Finally, Harper et al. (2002) emphasize that the goal of tribal exposure scenarios is to protect traditional lifeways not simply as they are currently practiced within the boundaries of tribal lands but also as they were historically practiced at the time treaty rights were established.

The introduction of tribal traditional lifeways scenarios corresponds to Latour's (2004) notion of perplexity, or "the number of propositions to be taken into account in the discussion" (p. 109). By *proposition*, Latour does not mean a linguistic statement but "an association of humans and nonhumans before it becomes a full-fledged member of the collective, an instituted essence" (p. 247). A proposition is a human–nonhuman association that a collective has not yet taken into account as part of the common world. During the 1980s, as quantitative risk assessment became the privileged technique for assessing environmental risks to public health in the United States, tribal traditional lifeways, as unique associations of humans and nonhumans, remained largely invisible. Because risk assessment failed to take them into account, they were, in effect, not yet full-fledged members of the collective. Indeed, historical processes of internal colonialism within the United States had explicitly attempted to exteriorize and even exterminate tribes' traditional ways of life, so it was no surprise that they remained invisible to regulatory science in the 1970s and 1980s (Deloria & Lytle, 1998). But by the end of the 1990s, a distinctive set of tribal publics to be protected—located and placed in very specific ways by internal US colonialism—henceforth needed to be taken into account.

Nonetheless, even while tribal risk assessors began insisting that EPA account for tribal traditional lifeways at risk, it became clear that the issue was more complicated than realized up to that point. With the assertion of tribal traditional lifeways, the matter was not simply one of how to democratize the sciences in a more appropriate way. It was also about the encounter between two distinct collectives, or ways of collecting the world: the Western or modern collective that posits nature and society as separate, and a nonmodern collective that insists on the inseparability of humans and the nonhuman environment. Traditional, non-Western ways of knowing the world play a more central role in the latter, and part of the demand to take traditional tribal lifeways into account was the requirement to recognize the legitimacy of traditional knowledge as well. This requirement has generated tension between

Western-trained risk assessors seeking to localize risk assessment so as to work more effectively in Indian Country and tribal members seeking to replace the risk assessment paradigm entirely. Moreover, it has broadened the debate to encompass not only the issue of how to make sciences public but also how to make them coexist credibly with the kinds of knowledge belonging to other collectives.

Consultation: The Tribal Science Council and the Risk Assessment Workshop

By the end of the 1990s, involvement in EPA's regulatory science activities had become an issue of paramount importance for tribal and intertribal agencies. In 1999 the National Tribal Caucus (tribal representatives within the National Tribal Operations Committee) proposed the formation of a Tribal Science Council "to provide a structure for Tribal involvement in the Agency's science efforts" (National EPA-Tribal Science Council, n.d.). The new council would give tribes a means to influence EPA's approach to regulatory science by determining the scientific issues of highest priority for tribes throughout the United States and by formally working with EPA to generate new science policies for Indian Country. The Tribal Science Council developed its initial membership and met for the first time near the end of 2001.

At a meeting in September 2002, the Council drew on input from tribes in the ten EPA regions to identify a preliminary set of science priorities (National EPA-Tribal Science Council, 2006a). At the top of the list was the question of tribal traditional lifeways, "including tribally relevant risk assessment and a new concept for environmental decision making" based on health and well-being instead of risk (p. 5). Although the Council established as a short-term goal the integration of tribal traditional lifeways into EPA's current approach to risk assessment, it set as a long-term goal the development of an entirely new paradigm for regulatory science based not on risks to individuals but on "human and ecological health and well-being" (National EPA-Tribal Science Council, 2005, p. 3). The establishment of goals with different time frames in part reflected an effort to resolve the tension between approaches to risk grounded in Western science and approaches grounded in traditional knowledge.

Over the next 2 years, addressing tribal traditional lifeways went from being the Tribal Science Council's (2005) "top science priority" among many to an "overarching issue" (p. 3): a framework that unified all of the other priorities and concerns. It became clear during two Tribal Science Council workshops in 2003—one on the topic of risk assessment and health and well-being and the other on the more general theme of health and well-being—that the matter of how to address tribal traditional lifeways would require an additional workshop (National EPA-Tribal Science Council, 2005). Although the workshop would to some degree address the long-term goal of developing an alternative to risk assessment, its primary focus would be the short-term goal of modifying conventional risk assessment so that it could take into account the distinctive exposures and risks associated with tribal traditional lifeways. The goals of the workshop would be modest: to "generate ideas," "develop recommendations," and "begin dialogue" that would lead up to formal consultation with individual tribes rather than create new science policy (p. 3).

The workshop, which took place in Sparks (a suburb next to Reno, Nevada) in January 2005, brought together a Recommendations Development Workgroup, made up of "invited tribal and EPA risk assessment practitioners and policy analysts" (National EPA-Tribal Science Council, 2005, p. 4), and a larger group of general "workshop participants." The Workgroup was an expert panel, with twelve representatives from tribes or intertribal organizations and five representatives from EPA's regions or program offices. Most of these individuals were risk assessment practitioners or had experience working with tribes on risk assessment. Most of the general workshop participants were tribal members, tribal agency representatives, representatives from Alaska Villages, or representatives from intertribal organizations. Others were affiliated with EPA, academic institutions, consultants, law firms, and environmental organizations.

At this workshop the matter of consultation emerged as a crucial issue. For Latour (2004), *consultation* pertains to the "number of voices that participate in the articulation of propositions" (p. 106). In Indian Country, the concept of consultation has distinctive connotations associated with the unique status of tribes as sovereign governments.[3] Although there is no standard definition of consultation, in the context of tribal law and policy it refers to formal dialogue between the federal government and tribal governments (National Environmental Justice Advisory Council Indigenous Peoples Subcommittee, 2000). Tribes emphasize that it is not the same as public participation, which refers to the rights of US citizens—or the general public—to be involved in federal decision-making processes. Although tribal members, as US citizens, are also members of the general public with rights to participation, tribal governments represent distinct publics or polities that are not reducible to subdivisions of the general public. To use Latour's vocabulary, the number of voices that participate in the articulation of propositions pertaining to Indian Country should not be limited to individual tribal members speaking for themselves but should also include representatives speaking for particular tribal publics.

The workshop revealed ways in which the appropriate localization of consultation itself—or the involvement of publics in articulating propositions—also plays a crucial role in developing credible risk assessments in Indian Country. The examples that follow briefly highlight two dimensions that became salient: (a) the places and scales for conducting consultation with tribes and tribal members and (b) the microgeographies or spatial arrangements within which this consultation should take place.

First, several of the general participants expressed their confusion over whether the workshop constituted a formal consultation between the EPA and tribes. Were workshop participants from tribes and Alaska Native villages speaking for themselves as tribal members but also as US citizens? Or did the expert workgroup panel understand them to be speaking for *tribes*—and, by extension, for the distinctive publics that tribes constitute? A number of indigenous participants emphasized that "they did not have proper authority from their Tribal Council or governing body to make formal recommendations to EPA," and they urged EPA to "enter into government-to-government consultation with their individual tribes" to develop such recommendations (National EPA-Tribal Science Council, 2005, p. 13). When one Workgroup panelist floated the idea of using the US government's centralized *Federal Register* as a mechanism for consulting with tribes, his suggestion met with strong objections. One participant recommended that EPA "send representatives out to all tribes in Indian Country to explain the issues, any draft documents developed by the Workgroup, and answer questions raised by tribes" (p. 14).

Workgroup members subsequently clarified that the workshop was not intended to be a formal consultation with tribes, but there was lingering uncertainty about exactly whom or what workshop participants were speaking for. There was general agreement that the voices of tribes—and, in particular, tribal elders—needed to be included more prominently in the development of risk assessments. Nevertheless, there was no consensus on the best way to consult with and seek input from tribes. It also became clear that, for many participants, an acceptable process of consultation would demand more than a nationally centralized workshop or notice. It would require EPA and the Tribal Science Council to send representatives out to the many different reservations and villages that make up Indian Country; consultation would need to be localized to the multiplicity of tribal publics.

A second issue that surfaced was the microgeography, or spatial arrangements, of the hotel ballroom where the workshop took place. The workshop's initial discussion, focused on the long-term objective of developing an alternative to the risk assessment paradigm, was open to both the Workgroup and the general participants. However, in the afternoon, as the topic of discussion shifted to the short-term objective of addressing tribal traditional lifeways within EPA's current approach to risk assessment, "[t]he workshop was to move from a more open format with full public participation to a "fish bowl" format, whereby discussion was to focus around workgroup members, with opportunities provided for public input and questions from workshop participants at regular intervals" (National EPA-Tribal Science Council, 2005, p. 10). General participants and Workgroup panelists alike expressed frustration about the "fish bowl" arrangement, in which panelists faced each other instead of the general participants. After meeting privately after the first day of the workshop, tribal participants explained that the arrangement of the ballroom reinforced a lay-expert divide that made the acceptance of recommendations less likely. "It was felt that there was a disconnect between the people who came to learn about risk assessment and the "experts"... People did not trust the fish bowl format/process" (p. 15). As a consequence, much of the second morning of the workshop was dedicated to changing its format and tone. Participants introduced themselves and shared their stories, and workshop staff "broke" the fish bowl by rearranging the tables and chairs. After the reconfiguration of the space, discussion on how to address tribal traditional lifeways in risk assessment finally resumed.

Subsequently, the perplexity introduced by tribal traditional lifeways at risk returned to the fore. Discussion came back to the question of how to localize the distinctive tribal publics whose health risk would be assessed. An environmental chemist on the panel alluded to the question of scale, indicating that "developing tribally specific models does not necessarily require the development of individual models for each tribe in Indian Country" but that instead scenarios based on ecoregions could potentially be used (National EPA-Tribal Science Council, 2005, p. 17). He pointed out that Barbara Harper, a risk assessor on the Workgroup panel who had collaborated on the first tribal traditional lifeways exposure scenarios (see above), was currently developing ecoregional scenarios that could be adjusted to individual sites in Indian Country more effectively than "existing suburban population default models" (p. 17). There was general consensus among panelists that addressing tribal traditional lifeways in risk assessment should not develop nationally-scaled exposure factors; such factors would not reflect regional and site-to-site variation in eating habits, resource use, and other important exposure variables. Although space does not permit further description of the workshop, it should be clear from these brief examples that the workshop was an event in which the geographic requirements for credibility and legitimacy surfaced in prominent ways.

After this third workshop, the Tribal Science Council extracted from the workshop series several specific requirements for improving the credibility of the risk assessment process within Indian Country (National EPA-Tribal Science Council, 2006b). Several of them pertained primarily to the matter of consultation, construed broadly: (a) increasing opportunities both for EPA to educate tribes about the risk assessment process and for tribes to educate EPA about tribal values, (b) consulting formally with tribes and ensuring their full inclusion in decision-making processes, and (c) making adequate funding available to enable tribes to participate meaningfully in the risk assessment process. In addition, the Council recognized that increasing the credibility of risk assessments in Indian Country would require transforming the process of data collection so as to (d) incorporate traditional tribal knowledge as a valid input to risk assessment; (e) integrate qualitative as well as quantitative data; (f) involve tribes in the development of sampling protocols, identification of exposure indicators, and other data collection activities; and (g) ensure that tribes can both protect confidential data and ensure that the data they gather are considered valid. In essence, the credibility of risk assessment in Indian Country would require localizing consultation not simply to the broader "tribal public"—that is, to American Indians as a national constituency—but to the specific and distinctive tribal publics of individual reservations and villages.

Other requirements articulated in the document pertained more to the perplexity associated with measuring risks to tribal publics, that is, the need to "take into account or allow for unique characteristics of tribes and tribal communities that create unique tribal exposures" (National EPA-Tribal Science Council, 2006b, p. 17). Several aspects contribute to the uniqueness of tribes as publics at risk: their small population size and, consequently, the vulnerability of their distinctive cultures; their ties to "fixed land and resource bases"; and their "unique dietary, cultural, and religious practices" (pp. 17–18). The document outlines three broad ways to localize risk assessment practices in order to address these unique exposures credibly. The first way is to expand risk assessment models to include more sensitive populations than the general public. The second way is to develop default exposure values and exposure scenarios for specific communities, or at least ecoregions, rather than for US tribes in general. The third broad way to localize risk assessment practices is to give equal weight in risk assessments to qualitative risks, such as impacts on tribal cultures. Again, although the Tribal Science Council reached no definitive conclusion about the appropriate scale of localization—community or ecoregion— it became clear that legitimate risk assessment in Indian Country should assess risks not to a nationally generalized tribal *public* but to locally specific tribal *publics*.

Conclusion

At the interface of science and public policy, the requirements of credibility and legitimacy take on distinctive geographic dimensions—particularly in the encounter with tribal lands and other spaces of difference. On the one hand, regulatory agencies may seek to shore up the credibility of their methods and techniques by standardizing them across the territorial space of the nation-state, effectively delocalizing them. EPA sought to do exactly that through its reforms of risk assessment in the 1990s. However, regulatory sciences differ from academic sciences in that securing credibility and legitimacy may also require finding appropriate ways to localize the same methods and techniques, both in the sense of customizing them to locally distinctive situations and concerns and in the sense of nurturing involvement by place-specific communities.

In the United States, Indian Country has been the most prominent setting where the requirement for EPA to localize its approach to human health risk assessment has emerged. It is one of the clearest examples of the multiplicity of publics with which regulatory science—and, indeed sciences in general—must engage. Tribal publics not only face unique exposures and risks, which call for local modifications to standardized risk assessment protocols and assumptions, but also hold distinctive rights and powers as "domestic dependent nations," which obligate EPA to go beyond ensuring the participation of the general public in decision-making. The effort to address tribal traditional lifeways in risk assessment foregrounds the requirements of both perplexity and consultation: the need for a democratized science to take into account both a larger number of human–nonhuman associations at risk and a larger number of voices that must participate in articulating them.

A theoretical implication of this chapter is that Latour's conceptualization of the work of the collective, though valuable, must integrate geography and spatiality more fully than has been the case thus far in order to be useful for empirical accounts of engagements between publics and the sciences. On the one hand, Latour's (2005) work on the process of "reassembling the social" brings him close to contemporary debates in geography. He clarifies his controversial conception of a "flat" ontology, insisting that scales and places are not pregiven units for the analyst to determine but the outcomes of actors "*scaling, spacing,* and *contextualizing* each other" (p. 184). In this sense he echoes geographers concerned with the social construction of scale and other geographic dimensions (Delaney & Leitner, 1997). However, Latour does not examine the integral roles that localizing plays in the work of the collective or in the more specific power of taking into account. One aim of this chapter has been to explore these roles.

Although this case has unresolved uncertainties pertaining to the appropriate localization of perplexity and consultation in risk assessment, some aspects of this localization have become apparent. Even though it is debatable whether tribal traditional lifeways exposure scenarios should be localized to particular communities or to broader ecoregional scales that encompass multiple tribes, the credibility of these scenarios clearly demands that they not be constructed as *national*-scale "tribal defaults." Furthermore, although it remains uncertain how and whether an alternative health and well-being paradigm can eventually replace risk assessment, any approach to representing the connections between environmental contamination and human health in Indian Country must consider the health and viability of individuals as well as of communities and cultural traditions. Finally, although the question of the appropriate places and scales at which tribal consultation should take place may remain open, the voices involved in consultation must include not only individual tribal members within the general public but also tribes as distinct polities—and must include traditional knowledge alongside Western sciences. The political work of resolving the remaining uncertainties will be an important process for future geographies of science to investigate.

Acknowledgment The author thanks Heike Jöns, Kristin Sziarto, and an anonymous reviewer for their insightful comments on earlier drafts. The research for this chapter was supported by an Association of American Geographers Travel Grant, along with fellowships from the US Environmental Protection Agency and the University of Minnesota Graduate School.

Notes

1. Indian Country is a legal term encompassing land within reservations, "dependent Indian communities," and allotments (Indian Country Crimes Act, 1948).
2. See http://www.epa.gov/superfund/programs/reforms/types/risk.htm
3. American Indian sovereignty, a notoriously ambiguous concept, is limited by the US Congress.

References

Biolsi, T. (2005). Imagined geographies: Sovereignty, indigenous space, and American Indian struggle. *American Ethnologist, 32*, 239–259.

Bowker, G. C., & Star, S. L. (1996). How things (actor-net) work: Classification, magic and the ubiquity of standards. *Philosophia, 25*, 195–220.

Delaney, D., & Leitner, H. (1997). The political construction of scale. *Political Geography, 16*, 93–97.

Deloria, V., & Lytle, C. M. (1998). *The nations within: The past and future of American Indian sovereignty.* Austin, TX: University of Texas Press.

EPA (Environmental Protection Agency) (1997). *Exposure factors handbook.* Washington, DC: US Environmental Protection Agency.

Frantz, K. (1999). *Indian reservations in the United States: Territory, sovereignty, and socioeconomic change*. Chicago: University of Chicago Press.

Harper, B. L., Flett, B., Harris, S., Abeyta, C., & Kirschner, F. (2002). The Spokane Tribe's multipathway subsistence exposure scenario and screening level RME. *Risk Analysis, 22,* 513–526.

Harris, S. G. (2000). Risk analysis: Changes needed from a native American perspective. *Human and Ecological Risk Assessment, 6,* 529–535.

Harris, S. G., & Harper, B. L. (1997). A native American exposure scenario. *Risk Analysis, 17,* 789–795.

Harris, S. G., & Harper, B. L. (1999, May 4). *How incorporating tribal information will enhance waste management decisions: US Department of Energy Long-Term Stewardship Information Center*. Retrieved August 30, 2006, from http://lts.apps.em.doe.gov/center/reports/doc16.html

Indian Country Crimes Act of 1948. 18 U.S.C. § 1151 (2004).

Jasanoff, S. (1986). *Risk management and political culture: A comparative study of science in the policy context*. New York: Russell Sage Foundation.

Jasanoff, S. (1990). *The fifth branch: Science advisers as policymakers*. Cambridge, MA: Harvard University Press.

Jasanoff, S. (1995). Procedural choices in regulatory science. *Technology in Society, 17,* 279–293.

Jasanoff, S. (Ed.). (2004). *States of knowledge: The co-production of science and social order*. London, New York: Routledge.

Kuehn, R. R. (1996). The environmental justice implications of quantitative risk assessment. *University of Illinois Law Review, 1,* 103–172.

Latour, B. (1987). *Science in action: How to follow scientists and engineers through society*. Cambridge, MA: Harvard University Press.

Latour, B. (1999). *Pandora's hope: Essays on the reality of science studies*. Cambridge, MA: Harvard University Press.

Latour, B. (2004). *The politics of nature: How to bring the sciences into democracy* (Catherine Porter, Trans.). Cambridge, MA: Harvard University Press.

Latour, B. (2005). *Reassembling the social: An introduction to actor-network-theory*. Oxford, England: Oxford University Press.

Law, J., & Mol, A. (2001). Situating technoscience: An inquiry into spatialities. *Environment and Planning D: Society & Space, 19,* 609–621.

Livingstone, D. N. (2003). *Putting science in its place: Geographies of scientific knowledge*. Chicago: University of Chicago Press.

National Environmental Justice Advisory Council Indigenous Peoples Subcommittee. (2000). *Guide on consultation and collaboration with Indian tribal governments and the public participation of indigenous groups and tribal members in environmental decision making*. Available at http://www.epa.gov/compliance/resources/publications/ej/nejac/ips-consultation-guide.pdf

National EPA-Tribal Science Council. (2005). *Addressing tribal traditional lifeways in EPA's Risk Assessment Policies and Procedures Workshop* (Workshop Summary). Reno/Sparks, Nevada, January 25–27.

National EPA-Tribal Science Council (2006a, April). *National tribal science priorities*. Washington, DC: US Environmental Protection Agency.

National EPA-Tribal Science Council. (2006b, April). Paper on tribal issues related to tribal traditional Lifeways, risk assessment, and health & well being: Documenting what we've heard. Washington, DC: US Environmental Protection Agency.

National EPA-Tribal Science Council. (n.d.). *National EPA-Tribal Science Council: A collaborative effort between Tribes and EPA to address priority environmental science issues* [Brochure]. n.p.: US Environmental Protection Agency.

Rouse, J. (1987). *Knowledge and power: Toward a political philosophy of science*. Ithaca, New York: Cornell University Press.

Silvern, S. E. (1999). Scales of justice: Law, American Indian treaty rights and the political construction of scale. *Political Geography, 18*, 639–668.

Wynne, B. (Ed). (1987). *Risk management and hazardous waste: Implementation and the dialectics of credibility*. Berlin, Heidelberg, New York: Springer-Verlag.

Abstract of the Contributions

Landscape of Knowledge

David N. Livingstone

Abstract The chapter begins with an overview of the "geography of scientific knowledge" project and ways that spatial questions have been shaping inquiries into both the production and circulation of science. The focus then shifts to four areas of research that can further deepen the enterprise and consolidate the value of a spatial perspective on knowledge. The first area is landscape agency. The author considers the part that landscape plays in the production and circulation of Darwin's theory of evolution by natural selection and argues for increasing the role accorded to the agency of landscape in understanding the development and circulation of Darwinism. In the second area, political ecology, discussion turns to the political shaping of landscape experience and thoughts on pre-Darwinian theories of human origins in order to explore the political geography of racial discourses about human beginnings. Addressing the third research area, print culture, the author notes that recent work on the geography of textuality is opening up new spheres of inquiry into knowledge circulation. He examines how various texts have been read in different geographical and cultural locations. His aim in considering the fourth research area, speech space, is to reflect on the connections between location and locution and thus the importance of attending to how scientific theories are talked about in different speech spaces.

Keywords Geography of scientific knowledge, Darwinism, Landscape agency, Political ecology, Print culture, Speech space

P. Meusburger et al. (eds.), *Geographies of Science*, Knowledge and Space 3,
DOI 10.1007/978-90-481-8611-2, © Springer Science+Business Media B.V. 2010

Global Knowledge?

Nico Stehr

Abstract The observations in this chapter are presented with a simple conceptual model of various sociological aspects relating to the idea of global knowledge. It deals not with an already existing worldwide community of knowledge but rather with the social and intellectual processes and obstacles that knowledge must master to achieve global scope and to overcome the unbalanced distribution of knowledge across societies. After skeptically enumerating allegedly preexisting worlds of knowledge and indicating where they are supposedly found, the author uses his concept to delimit the topic. Guided by a set of assumptions, he proceeds by discussing aspects that raise hopes that globalizing worlds of knowledge might exist. He then calls attention to social processes and to features of knowledge that make the imminent implementation of global worlds of knowledge seem rather unlikely. The chapter closes with a series of questions that necessarily remain open.

Keywords Knowledge production, Dissemination of knowledge, Global knowledge, science, Globalization, Theory

A Geohistorical Study of "The Rise of Modern Science": Mapping Scientific Practice Through Urban Networks, 1500–1900

Peter J. Taylor, Michael Hoyler, and David M. Evans

Abstract Using data on the "career" paths of one thousand "leading scientists" from 1450 to 1900, what is conventionally called *the rise of modern science* is mapped as a changing geography of scientific practice in urban networks. Four distinctive networks of scientific practice are identified. A primate network centered on Padua and central and northern Italy in the sixteenth century expands across the Alps to become a polycentric network in the seventeenth century, which, in turn, dissipates into a weak polycentric network in the eighteenth century. The nineteenth century marks a huge change of scale as a primate network centered on Berlin and dominated by German-speaking universities. These geographies are interpreted as core-producing processes in Wallerstein's modern world-system. The rise of modern scientific practice is central to the development of structures of knowledge that relate to, but do not mirror, material changes in the system.

Keywords Modern science, Space of flows, Scientists, Scientific centers, Scientific practice, Urban, Networks

From Mediocrity and Existential Crisis to Scientific Excellence: Heidelberg University Between 1803 and 1932

Peter Meusburger and Thomas Schuch

Abstract Since its founding in 1386, the University of Heidelberg has experienced ebbs and flows in its scientific prestige, intellectual influence, and attractiveness among scholars and students alike. It descended into mediocrity during Europe's religious dissention and international wars of the late 1600s to the early 1800s, then rose to become one of the world's leading research universities. Examining the changing career paths of this institution's professors, the author analyzes how it recovered its standing. He first discusses the factors that catapulted Heidelberg University from existential crisis into the top group of German universities within a few decades. He then analyzes the extent to which Heidelberg University's internal reorganization, improved academic standards, relatively favorable financial circumstances, growing scientific repute, and new policies on professorial appointments mirrored the social and regional changes and the career moves of the recruited academics. The focus is not on the individual biographies of a few eminent scholars but rather on the dynamic changes of social structures and on the spatial mobility of all Heidelberg professors between 1803 and 1932.

Keywords Academic career, Geography of science, Heidelberg University, Social origin of scholars, Spatial mobility

Academic Travel from Cambridge University and the Formation of Centres of Knowledge, 1885–1954

Heike Jöns

Abstract This chapter draws attention to academic travel as a key issue in the geographies of knowledge, science, and higher education. Building upon recent work in science studies and geography, the author argues that academic travel reveals the wider geography of scientific work and thus of the knowledge and networks involved. The study examines academic travel from Cambridge University between 1885 and 1954 to clarify the role of such movement in the development of Cambridge as a modern research university, the emergence of global knowledge centers elsewhere, and the development of an Anglo-American academic hegemony in the twentieth century. Using unpublished archival data on all recorded applications for leave of absence by Cambridge University Teaching Officers, the chapter also explores how the global geographies of academic travel varied among different types of work and thereby exposes distinct hierarchies of spaces of knowledge production and sites of study.

Keywords Academic travel, Knowledge production, Geographies of science, Transnational networks, Higher education, Cambridge University

Big Sciences, Open Networks, and Global Collecting in Early Museums

Dominik Collet

Abstract During the seventeenth century, many European collectors tried to establish their museums as global nodes of knowledge. To this end, they left their closed circles of learned men behind, and tapped into the commercial and colonial networks, relying on "weak ties" rather than personal acquaintance. In this "open network," objects and written correspondence traveled along different paths. Reuniting these references in the museum environment posed a serious challenge. This chapter follows the attempts made by the Royal Society of London at circulating references and constructing evidence. The ultimate failure of the fellows' ambitious plans highlights the hierarchies and inequalities inherent in open networks. The author thereby critically reflects on the role of commerce and weak ties in recent research and the significance of "space" for the study of early modern science.

Keywords Network theory, Early modern science, Natural history, Royal society, Collecting

Is the Atrium More Important than the Lab? Designer Buildings for New Cultures of Creativity

Albena Yaneva

Abstract The new challenge for architects of scientific buildings since the early 1990s has been to reinvent the potential of the spaces for collaborative research in order to increase their social performance. A new generation of signature architects and lab consultants has shifted the focus from labs to atriums. As a highly interconnective space, the atrium creates specific cognitive environments that facilitate networking and thereby stimulate new collaborative synergies. It also serves as a chamber for mixing human and nonhuman actors, often forming a complex quasi-urban network in large buildings shaped like cities. Both architects and scientists-as-clients have redefined their roles, designing spaces that do not determine the research conducted therein but instead mediate particular cognitive activities and diffuse them through complex networks. The new scientific design also redefines the entire cosmologies of the worlds of science.

Keywords Laboratory design, Atrium, Collaborative research, Cosmologies of science, Architecture of science

Outer Space of Science: A Video Ethnography of Reagency in Ghana

Wesley M. Shrum, Ricardo B. Duque, and Marcus Antonius Ynalvez

Abstract Current approaches to science and technology in developing areas depend excessively on an outmoded concept ("development") and fail to address spatiality in a way that can be adequately communicated. This chapter describes an audio-visual approach based on the concept of reagency, focusing on the problems of Internet connectivity in the scientific institutes of Ghana. The authors examine the reagency process through conflicts among collaborators who seek to implement a project with funds from abroad together with local agents who possess a variety of reputations and interests. The frustrations and apparent failure of the project are the obvious theme of a conventional account, but the experience is also examined through the eyes of one local collaborator who plans a successful return to his village. The authors argue that understanding geographies of science, particularly in the "outer space" of Africa, requires one to suspend judgment about the eventual outcomes of these processes and benefits from supplementing traditional approaches with video ethnography.

Keywords Internet connectivity, International collaboration, Science, Reagency, Video ethnography, Ghana

The Making of Geographies of Knowledge at World's Fairs: Morocco at Expo 2000 in Hanover

Alexa Färber

Abstract Until the mid-twentieth century, world fairs demonstrated national identity through industrial and cultural production and conveyed images of a world structured by hierarchical relations between nations, empires, and colonies; by the center and the periphery; and by cultures and knowledge. This chapter shifts attention to representational work, the geographies of knowledge operating at world fairs. The author seeks to make these geographies visible in their historical depth, ambivalence, and social relevance. The selected framework is Morocco's exhibit at "Expo

2000," the world's fair that took place in Hanover, Germany. The aim is to show how the restricted use of technology and the stress on creating an oriental atmosphere helped translate information and differentiated messages in the exhibit. The author argues that the "new smartness" (Andrew Ross) prevalent in the knowledge production in Morocco was voluntarily made invisible.

Keywords Representational work, Knowledge work, Technologies of nationhood, Knowledge society, World fair, Orientalism

Geographies of Science and Public Understanding? Exploring the Reception of the British Association for the Advancement of Science in Britain and in Ireland, c. 1845–1939

Charles W. J. Withers

Abstract The chapter examines evidence for the differential reception of science within different urban audiences following the meetings of the British Association for the Advancement of Science (BAAS) from the mid-nineteenth century on. Attention is paid to the professional reception of BAAS science, to the civic and political context of BAAS science in Britain and Ireland, to gender differences, and to varying reception of different subjects within BAAS meetings as broadly specialist or popular sciences. The chapter illustrates differences within the geography of science and reveals the science of geography to have been seen as a popular science by public audiences despite claims from disciplinary "professionals" over its specialist scientific status.

Keywords Audiences, British association for the advancement of science, Geography, Reception, Public understanding of science

Testing Times: Experimental Counter-Conduct in Interwar Germany

Alexander Vasudevan

Abstract This chapter builds on recent geographical approaches to the investigation of scientific experimentation. Although a number of studies have explored the various sites of scientific practice and the role of space in the constitution of experimental matters of fact, far less attention has been directed to the *cultural* *geographies* of experimental science and the extrascientific zones in which modes of experimental practice were themselves developed and contested. Drawing on the reception of professional psychiatry in interwar Berlin (1919–1933), this chapter traces an alternative set of "experimental systems" that seized on and countered the credibility of psychiatric expertise. It focuses, in particular, on a series of modernist experiments in interwar Germany that actively reconfigured psychiatric science as a series of critical aesthetic interventions themselves tasked with performing "scientific experiments."

Keywords Experimental practice, Cultural geographies of science, Modernism, Art and scientific practice, Weimar Germany, Dadaism, Experimental psychiatry, Bertolt Brecht, Epic theater

NGOs, the Science-Lay Dichotomy, and Hybrid Spaces of Environmental Knowledge

Sally Eden

Abstract This chapter considers how NGOs see and manage their place in relation to the science–lay dichotomy frequently invoked in modernist arguments about environmental science. The author draws on two empirical studies, one on general notions of science, expertise, and credibility among United Kingdom NGOs involved in waste debates, the other on environmental knowledge for certifying sustainable products by the Forest and Marine Stewardship Councils. She borrows Gieryn's (American Sociological Review, 48, 781–795, 1983; Handbook of science and technology studies, pp. 393–444, 1995; Cultural boundaries of science: Credibility on the line, 1999; Knowledge and space: Vol. 1. Clashes of knowledge: Orthodoxies and heterodoxies in science and religion, pp. 91–99, 2008) notion of "boundary-work" to argue that NGOs hybridize the science–lay dichotomy in different ways and reflexively seek—or even create—hybrid spaces of science and lay knowledge to advance their own agendas for environmental reform. Such hybrid spaces are alternative knowledge spaces, but they are not necessarily public ones, created as they are by invitation and being subject to clear lines of demarcation and governance. Recognizing and pursuing hybridity, though not a simple panacea for the modernist assumption of a problematic science-lay dichotomy, may thus help challenge outdated models of the place science has in society and the ways in which cultural cartographies of science are shaped and used.

Keywords Nongovernmental organizations, Boundary work, Lay-expert dichotomy, Hybrid spaces, Environmental science, Certification, Credibility

Regulatory Science and Risk Assessment in Indian Country: Taking Tribal Politics into Account

Ryan Holifield

Abstract Geographies of science show that the credibility of scientific methods often depends on their universalization and delocalization. In regulatory science, however, credibility may also depend on appropriate localization. This chapter examines efforts by tribes to require the U.S. Environmental Protection Agency to localize its risk assessment policies in order to take account of the distinctiveness of tribal publics. Tribal publics are distinct from the general public both because practicing tribal traditional lifeways presents unique public health risks and because tribal sovereignty establishes tribes as distinctive polities. The question of public participation in risk assessment is further complicated by the encounter between Western science and Traditional Environmental Knowledge. The author argues both that Bruno Latour's requirements of perplexity and consultation in the collective are valuable for conceptualizing the debate over traditional tribal lifeways at risk and that geography is integral to the realization of these requirements.

Keywords Regulatory science, Risk assessment, American Indian, Traditional tribal lifeways, Science and public policy, Localization

The Klaus Tschira Foundation

Physicist Dr. h.c. Klaus Tschira established the Klaus Tschira Foundation (KTF) in 1995 as a not-for-profit organization conceived to support research in informatics, the natural sciences, and mathematics and to foster public understanding of these sciences. Klaus Tschira's commitment to this objective was honored in 1999 with the "Deutscher Stifterpreis," the prize awarded by the National Association of German Foundations. Klaus Tschira is a cofounder of SAP AG in Walldorf, one of the world's leading companies in the software industry.

The KTF provides support mainly for research in applied informatics, the natural sciences, and mathematics and funds educational projects for students at public and private universities and schools. The resources are largely used for projects initiated by the foundation itself. It commissions research from institutions such as EML Research, founded by Klaus Tschira. The central goal of that organization for applied informatics is to develop new information-processing systems whose technology is perceived as userfriendly. In addition, the KTF invites applications for projects that are in line with the central concerns of the foundation.

The seat of the KTF is Villa Bosch in Heidelberg (Fig. 1), the former residence of Carl Bosch (1874–1940), the Nobel Prize Laureate for Chemistry. Carl Bosch, scientist, engineer, and businessman, joined BASF (Badische Anilin- & Soda-Fabrik) in 1899 as a chemist and became its CEO in 1919. In 1925 he was appointed CEO of the then newly created IG Farbenindustrie AG, and in 1935 he became chairman of the supervisory board of this chemical conglomerate. In 1937 Bosch was elected president of the Kaiser Wilhelm Gesellschaft (later renamed as the Max Planck Gesellschaft), the premier scientific society in Germany. Bosch's work combined chemical and technological knowledge at its best. Between 1908 and 1913, together with Paul Alwin Mittasch, he solved numerous problems in the industrial synthesis of ammonia, drawing on a process discovered earlier by Fritz Haber (Karlsruhe), who won the Nobel Prize for Chemistry in 1918. The Haber-Bosch process, as it is known, quickly became the most important method of producing ammonia—and remains so to this day. Bosch's research also influenced high-pressure synthesis of other substances. He was awarded the Nobel Prize for Chemistry in 1931, together with Friedrich Bergius.

In 1922 BASF erected a spacious country mansion and ancillary buildings in Heidelberg-Schlierbach for its CEO, Carl Bosch. The villa is situated in a small park on the hillside above the Neckar river and within walking distance from the famous Heidelberg Castle. As a fine example of the style and culture of the 1920s, Villa Bosch is considered one of the most beautiful buildings in Heidelberg and has been declared a protected cultural site. After World War II, it served as a domicile for high-ranking military staff of the United States Army. Thereafter, a local enterprise used the villa as its headquarters for several years. In 1967 Süddeutsche Rundfunk, a broadcasting company, established its Heidelberg studio there. Klaus Tschira bought Villa Bosch as a future home for his planned foundations toward the end of 1994 and had the building restored and modernized. Combining the historic ambience of the 1920s with the latest infrastructure and technology, Villa Bosch reopened in new splendor in mid-1997, ready for fresh challenges. The former garage, located 300 m west of the villa, now houses the Carl Bosch Museum Heidelberg, founded and managed by Gerda Tschira and dedicated to the memory of the Nobel laureate, his life, and his achievements.

This book is the result of a symposium entitled "Geographies of Science," which took place at Villa Bosch, June 27–30, 2007 (Fig. 2).

For further information contact:

Klaus Tschira Foundation gGmbH
Villa Bosch
Schloss-Wolfsbrunnenweg 33
D-69118 Heidelberg, Germany
Tel.: (+49) 6221-533-101
Fax: (+49) 6221-533-199
beate.spiegel@ktf.villa-bosch.de

Public relations:
Renate Ries
Tel.: (+49) 6221-533-214
Fax: (+49) 6221-533-198
renate.ries@ktf.villa-bosch.de

http://www.klaus-tschira-stiftung.de

Fig. 1 Villa Bosch (© Peter Meusburger, Heidelberg)

Fig. 2 Participants of the symposium "Geographies of Science" at Villa Bosch in Heidelberg, June 27–30, 2007 (© Thomas Bonn, Heidelberg)

Index